Advances in Intelligent Systems and Computing

Volume 376

Series editor
Janusz Kacprzyk, Polish Academy of Sciences, Warsaw, Poland
e-mail: kacprzyk@ibspan.waw.pl

About this Series

The series "Advances in Intelligent Systems and Computing" contains publications on theory, applications, and design methods of Intelligent Systems and Intelligent Computing. Virtually all disciplines such as engineering, natural sciences, computer and information science, ICT, economics, business, e-commerce, environment, healthcare, life science are covered. The list of topics spans all the areas of modern intelligent systems and computing.

The publications within "Advances in Intelligent Systems and Computing" are primarily textbooks and proceedings of important conferences, symposia and congresses. They cover significant recent developments in the field, both of a foundational and applicable character. An important characteristic feature of the series is the short publication time and world-wide distribution. This permits a rapid and broad dissemination of research results.

More information about this series at http://www.springer.com/series/11156

Amr Mohamed · Paulo Novais
António Pereira · Gabriel Villarrubia González
Antonio Fernández-Caballero
Editors

Ambient Intelligence - Software and Applications

6th International Symposium on Ambient Intelligence (ISAmI 2015)

Editors
Amr Mohamed
Computer Science and Engineering
Department, College of Engineering
Qatar University
Doha
Qatar

Paulo Novais
Departamento de Informatica, ALGORITMI
Centre
University of Minho
Braga
Portugal

António Pereira
Escola Superior de Tecnologia e Gestão de
Leiria
Instituto Politécnico de Leiria
Leiria
Portugal

Gabriel Villarrubia González
Departamento de Informática y Automática
University of Salamanca
Salamanca
Spain

Antonio Fernández-Caballero
Departamento de Sistemas Informáticos
University of Castilla-La Mancha
Albacete
Spain

ISSN 2194-5357 ISSN 2194-5365 (electronic)
Advances in Intelligent Systems and Computing
ISBN 978-3-319-19694-7 ISBN 978-3-319-19695-4 (eBook)
DOI 10.1007/978-3-319-19695-4

Library of Congress Control Number: 2015940728

Springer Cham Heidelberg New York Dordrecht London

Printed on acid-free paper

Springer International Publishing AG Switzerland is part of Springer Science+Business Media
(www.springer.com)

Preface

This volume contains the proceedings of the 6th International Symposium on Ambient Intelligence (ISAmI 2015). The symposium was held in Salamanca, Spain during June 3–5 at the University of Salamanca.

ISAmI has been running annually and aiming to bring together researchers from various disciplines that constitute the scientific field of Ambient Intelligence to present and discuss the latest results, new ideas, projects and lessons learned, namely in terms of software and applications, and aims to bring together researchers from various disciplines that are interested in all aspects of this area.

Ambient Intelligence is a recent paradigm emerging from Artificial Intelligence, where computers are used as proactive tools assisting people with their day-to-day activities, making everyone's life more comfortable.

After a careful review, 27 papers from 10 different countries were selected to be presented in ISAmI 2015 at the conference and published in the proceedings. Each paper has been reviewed by, at least, three different reviewers, from an international committee composed of 74 members from 24 countries.

Acknowledgments

Special thanks to the editors of the workshops AIfES. Special Session on Ambient Intelligence for e-Healthcare.

We want to thank all the sponsors of ISAmI'15: IEEE Sección España, CNRS, AFIA, AEPIA, APPIA, AI*IA, and Junta de Castilla y León.

ISAmI would not have been possible without an active Program Committee. We would like to thank all the members for their time and useful comments and recommendations.

We would also like to thank all the contributing authors and the Local Organizing Committee for their hard and highly valuable work.

Your work was essential to the success of ISAmI 2015.

June 2015

Amr Mohamed
Paulo Novais
António Pereira
Gabriel Villarrubia González
Antonio Fernández-Caballero

Organization

Program Committee Chairs

Amr Mohamed (Chair)—College of Engineering, Qatar University (Qatar)
António Pereira (Chair)—Instituto Politécnico de Leiria (Portugal)

Program Committee

Adolfo Barroso Gallego—University of Salamanca
Alberto Lopez Barriuso—University of Salamanca
Alvaro Castro González—Universidad Carlos III de Madrid
Andreas Riener—Johannes Kepler University Linz, Institute for Pervasive Computing
Andrei Olaru—University Politehnica of Bucharest
Ângelo Costa—Universidade do Minho
Ansar-Ul-Haque Yasar—Universiteit Hasselt—IMOB
Antonio Fernández-Caballero—Universidad de Castilla-La Mancha
António Pereira—Escola Superior de Tecnologia e Gestão do IPLeiria
Balbo Flavien—ISCOD/Henri Fayol Institute
Bartolomeo Sapio—Fondazione Ugo Bordoni
Bogdan Kwolek—AGH University of Science and Technology
Carlos Ramos—Instituto Superior de Engenharia do Porto
Cecilio Angulo—Universitat Politcnica de Catalunya
Cesar Analide—University of Minho
Davide Carneiro—University of Minho
Eduardo Dias—CITI—FCT/UNL
Emmanuel Adam—University of Lille Nord de France
Enzo Pasquale Scilingo—University of Pisa
Fernando Silva—Department of Informatics Engineering; School of Technology and Management; Polytechnic Institute of Leiria, Portugal

Flavia Delicato—Federal University of Rio de Janeiro (UFRJ)
Florentino Fdez-Riverola—University of Vigo
Gabriel Villarrubia González—University of Salamanca
German Nemiorvskij—Albstadt-Sigmaringen University
Goreti Marreiros—ISEP/IPP-GECAD
Guillaume Lopez—Aoyama Gakuin University, College of Science and Technology
Habib M. Fardoun—King Abdulaziz University
Hans W. Guesgen—Massey University
Ilias Maglogiannis—University of Piraeus
Jaderick Pabico—University of the Philippines Los Banos
Javier Bajo—Universidad Politécnica de Madrid
Javier Carbo—Universidad Carlos III of Madrid
Javier Jaen—Polytechnic University of Valencia
Jean-Paul Jamont—LCIS, Université de Grenoble
Jesus Favela—CICESE
Jose Carlos Castillo Montoya—Universidad Carlos III de Madrid
Jose M. Molina—Universidad Carlos III de Madrid
José Machado—University of Minho, Computer Science and Technology Centre
Jose Manuel Pastor Garcia—Universidad de Castilla-La Mancha
José Ramón Álvarez-Sánchez—Universidad Nacional de Educación a Distancia
Joyca Lacroix—Philips Research
Juan Pavón—Universidad Complutense de Madrid
Julie Doyle—CASALA, Dundalk Institute of Technology
Junzhong Gu—East China Normal University
Koen van Dam—Imperial College London
Latif Ladid—UL & IPv6 Forum—IOT subC
Lawrence Wai-Choong Wong—National University of Singapore
Lourdes Borrajo—University of Vigo
Maria Navarro—University of Salamanca
Miguel A. Salichs—Universidad Carlos III de Madrid
Nadia Berthouze—University College London
Nuno M. Garcia—University of Beira Interior
Pablo Chamoso—University of Salamanca
Pascual González—Universidad de Castilla-La Mancha
Paulo Novais—University of Minho
Radu-Daniel Vatavu—University Stefan cel Mare of Suceava
Rafael Martinez Tomas—Universidad Nacional de Educación a Distancia
Remy Courdier—LIM—Université de la Réunion
Rene Meier—Lucerne University of Applied Sciences
Ricardo Costa—ESTGF.IPP
Rui José—University of Minho
S. Shyam Sundar—Penn State University, USA & Sungkyunkwan University, Korea
Sergio Sanchez González—University of Salamanca

Sotiris Nikoletseas—University of Patras and CTI
Soumya Kanti Datta—EURECOM
Stéphanie Combettes—IRIT—University of Toulouse
Sylvie Ratté—École de technologie supérieure
Teresa Romão—DI/FCT/UNL
Tibor Bosse—Vrije Universiteit Amsterdam
Valérie Camps—University of Toulouse—IRIT, France
Vicente Julian—GTI-IA DSIC UPV
Yi Fang—Purdue University

Organizing Committee Chairs

Paulo Novais—University of Minho (Portugal)
Gabriel Villarrubia González—University of Salamanca (Spain)

Local Organization Committee

Javier Bajo—Polytechnic University of Madrid (Spain)
Juan F. De Paz—University of Salamanca (Spain)
Fernando De la Prieta—University of Salamanca (Spain)
Sara Rodriguez—University of Salamanca (Spain)
Tiancheng Li—University of Salamanca (Spain)
Javier Prieto Tejedor—University of Salamanca (Spain)
Cesar Analide—University of Minho (Portugal)

Workshops

AIfeH.Special Session on Ambient Intelligence for e-Healthcare

Organizing Committee
Antonio Fernández-Caballero, University of Castilla-La Mancha, Spain
Pascual González, University of Castilla-La Mancha, Spain
Elena Navarro, University of Castilla-La Mancha, Spain

Contents

Using Evolutionary Algorithms to Personalize Controllers in Ambient Intelligence

Shu Gao and Mark Hoogendoorn

Abstract As users can have greatly different preferences, the personalization of ambient devices is of utmost importance. Several approaches have been proposed to establish such a personalization in the form of machine learning or more dedicated knowledge-driven learning approaches. Despite its huge successes in optimization, evolutionary algorithms (EAs) have not been studied a lot in this context, mostly because it is known to be a slow learner. Currently however, quite fast EA based optimizers exist. In this paper, we investigate the suitability of EAs for ambient intelligence.

Keywords Ambient intelligence · Evolutionary algorithms · Personalization · CMA-ES

1 Introduction

The rise of ambient intelligence is becoming more and more apparent in our daily lives: an increasing number of devices is surrounding us that perform all kinds of measurements and try to utilize this information in an intelligent way, for instance by controlling certain actuators or providing some form of feedback. In order for environments or devices to act sufficiently intelligent they need to be able to learn from the behavior of the user. Users can for instance have completely different preferences from each other, and hence, if only a single strategy would be deployed the system would never be effective and the user experience would be disappointing. In addition, devices need to learn how to cooperate with each other, and given the wealth of different devices on the market you cannot predefine the way in which they should.

S. Gao (✉) · M. Hoogendoorn
Department of Computer Science, VU University Amsterdam,
De Boelelaan 1081, 1081 HV Amsterdam, The Netherlands
e-mail: s.gao@vu.nl

M. Hoogendoorn
e-mail: m.hoogendoorn@vu.nl

A. Mohamed et al. (eds.), *Ambient Intelligence - Software and Applications*,
Advances in Intelligent Systems and Computing 376,
DOI 10.1007/978-3-319-19695-4_1

Learning of preferences and learning how to establish effective cooperation between devices has been a subject of study in the field of ambient intelligence (or under its closely related fields such as pervasive computing and ubiquitous computing), see e.g. [1, 10, 15]. In [1] three stages of adaptation are identified: (1) the initial phase during which data is collected; (2) learning of behavior based upon the data collected, and (3) coping with dynamic environments. Mainly in the first stage hardly any machine learning approaches are appropriate as they hardly have any data/experiences to learn from, whereas this is a crucial phase. In that phase, the learning algorithm should learn on-the-fly. One of the problem solvers known to work well in nature, evolutionary algorithms (EAs), has not received a lot of attention in this domain, and in particular not for the subproblem which has just been described. Although EAs are mostly seen as slow optimizers, they have been shown to work very well for a range of optimization problems, see e.g. [6]. Furthermore, approaches such as genetic programming (cf. [2]) are highly suitable to generate sophisticated controllers.

In this paper, we explore the suitability of EAs for an ambient intelligence task thereby assuming no data being available up front. More precisely, we study a scenario where multiple (possibly heterogenous) devices need to be controlled in a simple way, thereby taking the preferences of multiple users into account. The rationale for starting with a simple scenario is that we want to explore whether EAs are able to solve a relatively simple problem in a suitable way before we move on to more complex problems. We use the state-of-the-art evolutionary optimizer, namely the CMA-ES [9]. Given the nature of devices in ambient intelligence, we use different variants of the algorithm: a centralized versus representing a single central controller and a number of distributed controllers repressing individual devices with their own controller. As evaluation criteria we measure the quality of the solutions found in terms of the percentage from the optimal solution as well as the time required to find a reasonable solution. We compare the outcome with simple benchmark algorithms such as hill climbing and simulated annealing.

This paper is organized as follows: first, we present related work in Sect. 2. Thereafter, in Sect. 3 we present the learning approach and the experimental setup is presented in Sect. 4. The results are presented and analyzed in Sect. 5. Finally, Sect. 6 concludes the paper.

2 Related Work

As said in the introduction, a lot of authors acknowledge the importance of machine learning techniques in ambient intelligence. An overview of useful techniques as well as examples of machine learning applications are given in [1]. In quite some approaches, a dataset for training is assumed. For example, Mozer et al. [12] use artificial neural networks in an AmI environment. Classification is implemented in an environment named 'SmartOffice' by Gal et al. [7]. On the other hand, reinforcement learning is another approach that does not need training data which is applied to ambient intelligence environment by Mozer [11]. There are examples in which EAs are applied in Ambient Intelligence. Doctor, Hagras and Caalghan [5] for example use Genetic Programming as a benchmark algorithm, stating that GPs are less suitable to use in an online fashion as they require many generations. A Genetic Algorithm is applied in [3] but again not in an online fashion. In [4] EAs are used to compose software around applications. Hence, one can see that there is some work which combines EAs with Ambient Intelligence, but none have judged whether such approach could be suitable to use in an online fashion where users provide feedback and act as a fitness function.

3 Approach

In our approach, we assume an environment in which multiple ambient devices are present that are equipped with sensors and actuators and have controllers that express their behavior, i.e. map sensory values to actions. The mapping between devices and controllers is left open: on the one extreme each device could have its own controller whereas on the other side of the spectrum there could be a single controller for all devices jointly. In the environment one or multiple users are present each having their own preference in particular situations. Here, a situation is a unique combination of sensory values or possibly a set of such combinations which all map to the same situation. Learning such a mapping could be another learning endeavor but in this initial exploration of EAs for ambient intelligence this is beyond our scope. The main goal of our research is to create controllers for the ambient devices that satisfy the user preferences best, a problem which we formulate as a maximization problem of the following function:

$$F = \sum_{\forall S:SIT} \sum_{\forall U:USER} user\_satisfaction(U, S, actions_{controllers}(S, U))$$

In other words, the controllers should, for all situations, find the set of actions that satisfy the users most. Since the user satisfaction can only be provided by the user itself, this entails that the user needs to be consulted every time a new controller is generated. A secondary goal is therefore also to minimize the number of evaluations required by the algorithm to find the solution to avoid bothering the user too much, and the user having to bare a lot of non-satisfactory solutions.

We assume that the controller is optimized for each situation separately. For each of such situations, a controller is represented by means of numerical values for each action it can perform. For binary actions, the possible values are clearly limited to 0 and 1 whereas for continuous actions (e.g. light intensity, sound volume) the action can take any value which is appropriate for the action. Table 1 shows an example of such a representation.

Table 1 Example representation for one controller for a single situation

a_1	a_2	a_3	a_4
1	0.5	0	0.25

In order to solve the problem we have now created, we use a variety of different approaches: state of the art EAs, including the CMA-ES and Cooperative Co-Evolution as well as benchmark algorithms including hill climbing and simulated annealing. An alternative would also be to use reinforcement learning, but given the scope of the scenario explored in this paper (see Sect. 4), this is not a suitable option and therefore not discussed in detail here. Each of these algorithms is explained in more detail below.

3.1 CMA-ES

The CMA-ES [9] is an evolutionary strategy. In general, EAs work with a population of individuals (in our case expressing the value for actions for a certain situation), of which individuals are selected for mutation (on a single individual) and crossover (combining two or more individuals), resulting in new individuals. Out of the total pool of individuals a new population is selected again, and this process continues until some termination criterion is reached. An evolutionary strategy is a variant in which the individuals are composed of a series of real numbers and the individuals also contain a dedicated field which determines the mutation probability, this field is also subject to evolution, and hence, the mutation probability self-adapts. In the CMA-ES a covariance matrix is used in order to improve the effectiveness of generating new individuals. Explaining all details of the algorithm is beyond the scope of this paper, the reader is referred to [9] for more details. We use the CMA-ES in two variants: (1) a single evolutionary loop in which a central controller simply determines the action for the actuators of all devices for this particular situation, and (2) a decentralized approach where each device has its own controller and CMA-ES population to evolve the controller. For the latter case, we use the cooperative co-evolutionary approach as proposed by Potter and De Jong [13]. Here, a single device is selected while fixing the controllers of the other devices to the best one found until then. Each of the individuals of the population of the selected device is then evaluation in conjunction with these best controllers, resulting in a fitness score. After that, the next device is selected, etc. Devices are selected in a round robin fashion. For non-continuous actions the real value is rounded to the nearest value (i.e. 0/1) during the evaluation phase.

3.2 Standard GA

Next to the CMA-ES we also try a simpler variant of an EA, namely the so-called "standard GA", which, contrary to the sophisticated CMA-ES, consists of individuals that are composed of bits and uses less sophisticated operators. Combinations of bits can represent continuous actions for our case. Mutation takes place via simple bit-flips whereas crossover is done via selecting a crossover point and selecting the first part of one parent and the second part of the other. Selection is done by means of probabilities proportional to the fitness of the individual.

3.3 Hill Climbing and Simulated Annealing

For hill climbing we simply try a random step in either direction of the value for an action and perform an evaluation, the best solution (current, with the random step added, or deducted) is selected as the next controller. The process ends once the stopping condition has been met.

In the simulated annealing approach (see [14]), which works with the notion of the temperature of the process, a step is performed in the search space (i.e. for the action) which is equal to the temperature. The temperature function used is temperature function: $T = \frac{T_0}{log(k)}$ where T_0 is the initial temperature and k is number of steps. New solutions are accepted when they are an improvement, or, if not, they are selected with a probability $e^{\frac{\Delta C}{T}}$ which is dependent on the temperature of the process and the improvement made, ΔC. This scheme enables more exploration in the initial phase and more exploitation in the end of the process.

4 Experimental Setup

In this section, we describe the setup we have used to evaluate the algorithms that have been specified in Sect. 3. First, the case study is explained, followed by the precise setup of the experiments.

4.1 Case Study

As our case study, we focus on an office setting with lights that need to be controlled. As said, the focus is not so much on a complex scenario, but to study the potential of EAs for a relatively simple scenario. Figure 1 shows the scenario is more detail.

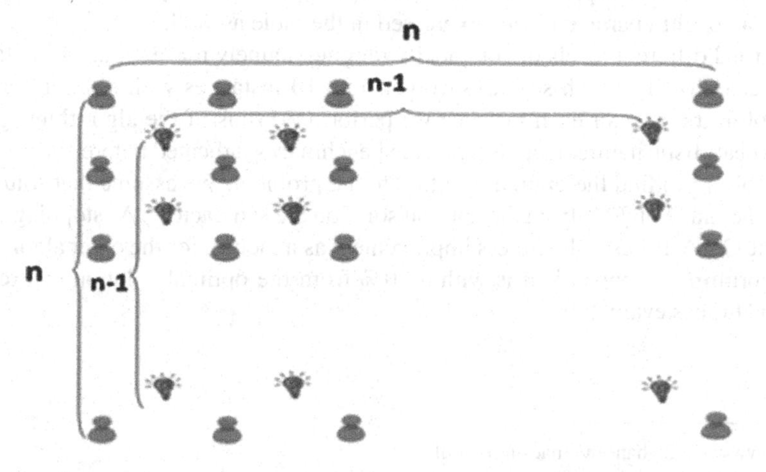

Fig. 1 Specific scenario

Essentially, there are more lights than users, and each user has a specific preference for a light intensity. The parameter n determines the complexity of the situation. Here, n defines the number of users (n^2) and the number of lights ($(n-1)^2$). We define such a complexity parameter as we want to study how the various approaches scale up with increasing complexity. We have chosen to have a number of lights which is smaller than the number of users to make the scenario more interesting. The intensity experienced by a user U is determined by all lights jointly, whereby the contribution of each individual light is determined by means of the following equation:

$$I(U) = \sum_{\forall L:LIGHTS} \frac{P(L)}{4\pi D(U,L)^2}$$

Here, *P(L)* is the power of the light L and *D(U, L)* is the distance of the light L to the user U. For now, we assume a single situation in which all users are present in the office. Each user has a preferred light intensity, thereby defining the function $user\_satisfaction(U, S, actions_{controllers}(S, U))$ as specified in Sect. 3:

$$user\_satisfaction(U, S, actions_{controllers}(S, U)) = |I(U) - pref\_int(U)|$$

4.2 Setup

We have implemented the entire system in Matlab, except of the CMA-ES which is available in C.[1] In our experiments, we run a number of different setups of the algorithm, in line with the approach outlined before, and which are shown in Table 2. CMA-ES is run with both a centralized and distributed controller setting, the standard GA only with a distributed setting, and the other benchmarks are only run in a centralized way. Note that the scenario does not contain any input states at the moment, making approaches such as reinforcement learning inappropriate. The precise algorithm parameters are expressed in the table as well.

We tried different levels of complexity ranging, namely n = 3,4,5,....,12, totaling to 10 scenarios. For each scenario we generate 10 instances with different preferences of users, and for each instance we perform 30 runs of the algorithms, given their probabilistic nature. In addition to the benchmarks indicated before, we also run an LP solver[2] to find the optimal solution to the problem. We assume that solutions within the range of 20 % from the optimal solution are satisfactory. As stopping criterion, the CMA-ES uses the fitness improvement as a metric, for the other algorithms the algorithm is stopped if it is within 20 % from the optimal solution or exceeds 100,000 fitness evaluations.

[1] https://www.lri.fr/~hansen/cmaesintro.html.

[2] http://lpsolve.sourceforge.net/5.5/.

Table 2 Experimental setup and settings

Abbreviation	Algorithm	Controller type	Specific settings
C-ES	CMA-ES	Centralized	Off-the-shelf toolkit, with population size set to $4 + (3 * log((n - 1)^2))$
D-ES	CMA-ES	Distributed	See above, population size is set to 4 for each controller
D-GA	Standard GA	Distributed	Number of bits: 16 Population size: 100 Crossover rate: 0.6 Mutation rate: $\frac{1}{16}$
C-SA	Simulated annealing	Centralized	$T_0 = 100$
C-HC	Hill climbing	Centralized	Random number is selected from the range [0, 10]

5 Experimental Results

The experimental results are described in this section. First, we look at the number of evaluations needed to come to a reasonable solution (i.e. within 20 % from the optimal value). Figure 2 shows the results for the various algorithms. From the graph, it can be seen that a centralized controller generated by the CMA-ES algorithm by far outperforms the alternative algorithms, although the performance of the decentralized CMA-ES variant is still relatively close. The scaling of the algorithm seems good, given the exponential nature of the number of lights that needs to be controlled as a function of n on the x-axis. To be more precise, for the simple scenario, including 9 users and 4 lights, the system could find good solution within 200 evaluations. But it needs over 1500 evaluations in a complex scenario which involves 121 lights and 144 users. So, although from a scientific perspective the speedup is good, from a user perspective it is quite cumbersome. The variation of performance between the different runs is low for the CMA-ES. When we look at hill climbing and simulated annealing, we see that hill climbing performs a lot worse, with a huge variation. Simulated annealing does better, but does not come close to the speed of the CMA-ES variants. The Standard GA in the distributed setting is worst, most likely due to the distributed setting in combination with the simplicity of the EA. Table 3 shows the complete overview of the average times to find a solution with 20 % from optimal.

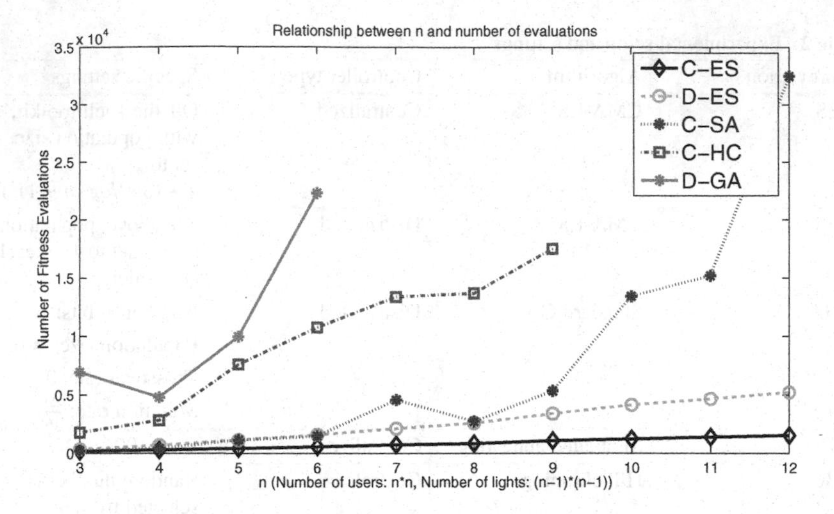

Fig. 2 Average time to optimal + 20 % for varying n

Furthermore, Fig. 3 shows the learning curve of the centralized CMA-ES for n = 3, it can be seen that the algorithm learns quite fast in the beginning, so the users are not exposed to very low quality solutions for a long period of time.

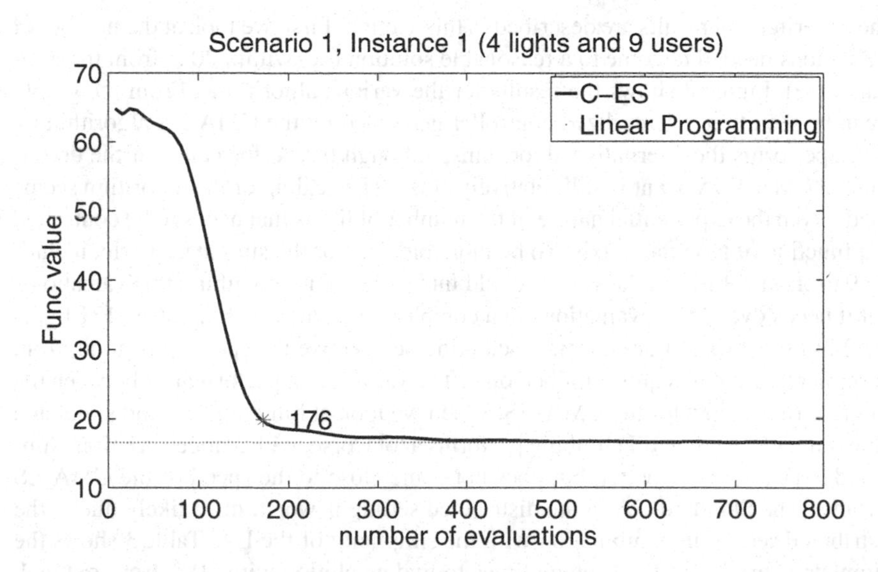

Fig. 3 The number of evaluations versus user satisfaction for C-ES

Table 3 Overview of the mean times to find a solution

Appr.	C-ES		D-ES		C-SA		C-HC		D-GA	
n	Ave	SD	Ave	SD	Ave	SD	Ave	SD	Ave	SD
3	179.2	28.9	333.6	54.3	206.8	107	1775.6	874.7	6941.8	10175.1
4	291	71.8	701.6	229.8	423	357.2	2828.3	2501.1	4806	3253.8
5	429.6	134.1	1148.8	443.7	1151	1789	7623.5	6169.5	9986.7	12361
6	569.4	87.9	1583.6	280.6	1484.3	919.2	10775.6	9118.9	22328.2	7644.7
7	743.4	206.8	2123.2	376.6	4580.3	7096.4	13388.7	12512.9	-	-
8	834	128.5	2584.4	352.7	2736.1	1572.1	13705.6	11093.5	-	-
9	1086.4	139.4	3396.8	350.3	5340.2	2325.6	17512.2	11957.3	-	-
10	1225.7	190.9	4140	282.8	13449.2	21352.6	-	-	-	-
11	1394	184.7	4648	401.6	15202.1	13940.4	-	-	-	-
12	1535.4	157.8	5206.8	552.9	32472.1	55847.1	-	-	-	-

6 Discussion

In this paper, we have explored the usability of EAs for personalization in ambient intelligence. Hereto, we have tried to formalize a fitness function, required for EAs, and have selected a first set of appropriate EA variants. In an experimental setting we have seen that EAs are able to find decent quality solutions, but as the problem becomes more complex the performance becomes a lot worse. Until now, the user feedback has just been the general level of satisfaction. Of course, more detailed feedback, or an initial phase of exploration could help to improve the speed to come to a solution and the quality of the solution. Ample approaches have utilized initial observations of users to derive a first set of reasonable controllers (see e.g. [8]). This was however not the purpose of this paper, we simply wanted to see whether an EA learning approach with one single piece of feedback could do the job, and the answer is that for simple environments this is possible, but as things get more complex this would become too much of a burden for users, let alone if multiple situations would need to be taken into account. Of course, the approach can still be applied, but our intuition is that one would need to resolve to alternative algorithms such as reinforcement learning.

For now an explicit fitness function in the form of user feedback has been obtained. We could also replace this with an alternative fitness function which is less direct (e.g. measure the work productivity), this would not change the setup of the learning system which shows how generic the approach is. How well the approach would learn the optimal lighting however would need to be studied, this would be an interesting aspect for future work. In addition, we want to explore more complex scenarios where sensors play a more prominent role and compare faster learning algorithms such as reinforcement learning to more knowledge driven approaches.

Acknowledgments This research is supported in part by scholarship from China Scholarship Council under number 201304910373. Furthermore, we would like to thank Gusz Eiben for the fruitful discussions and the anonymous reviewers for they valuable comments that helped to improve the paper.

References

1. A. Aztiria, A. Izaguirre, J.C. Augusto, Learning patterns in ambient intelligence environments: a survey. Artif. Intell. Rev. **34**(1), 35–51 (2010)
2. W. Banzhaf, P. Nordin, R. Keller, F. Francone, *Genetic Programming: An Introduction* (Morgan Kaufmann, San Francisco, 1998)
3. C.-K. Chiou, J.C. Tseng, G.-J. Hwang, S. Heller, An adaptive navigation support system for conducting context-aware ubiquitous learning in museums. Comput. Educ. **55**(2), 834–845 (2010)
4. O. Davidyuk, I. Selek, J. Imanol Duran, J. Riekki. Algorithms for composing pervasive applications. Int. J. Softw. Eng. Appl. **2**(2), 71–94 (2008)

5. F. Doctor, H. Hagras, V. Callaghan, A fuzzy embedded agent-based approach for realizing ambient intelligence in intelligent inhabited environments. IEEE Trans. Syst. Man Cybern. Part A Syst. Hum. **35**(1), 55–65 (2005)
6. A.E. Eiben, J. Smith, *Introduction to Evolutionary Computing* (Springer, London, 2003)
7. C.L. Gal, J. Martin, A. Lux, J.L. Crowley, Smart office: design of an intelligent environment. IEEE Intell. Syst. **16**(4), 60–66 (2001)
8. V. Guivarch, V. Camps, A. Peninou, Context Awareness in Ambient Systems by an Adaptive Multi-agent Approach, in *Ambient Intelligence*, vol. 7683, Lecture Notes in Computer Science, ed. by F. Paterno, B. de Ruyter, P. Markopoulos, C. Santoro, E. van Loenen, K. Luyten (Springer, Berlin, 2012), pp. 129–144
9. N. Hansen, A. Ostermeier, Adapting arbitrary normal mutation distributions in evolution strategies: the covariance matrix adaptation, in *Proceedings of IEEE International Conference on Evolutionary Computation*, May 1996, pp. 312–317
10. F. Mattern, M. Naghshineh, in *Pervasive Computing (Proceedings of the Pervasive 2002 International Conference)*, number 2414 in LNCS. Springer, 2002. Online version of the proceedings at http://link.springer.de/link/service/series/0558/tocs/t2414.htm
11. M.C. Mozer, *Lessons from an Adaptive Home* (Wiley, New York, 2005), pp. 271–294
12. M.C. Mozer, R.H. Dodier, M. Anderson, L. Vidmar, R. F.C. III, D. Miller. The neural network house: an overview (1995)
13. M.A. Potter, K.A.D. Jong, A cooperative coevolutionary approach to function optimization, in *Proceedings of the International Conference on Evolutionary Computation. The Third Conference on Parallel Problem Solving from Nature: Parallel Problem Solving from Nature*, PPSN III (Springer, London, UK, 1994), pp. 249–257
14. P. van Laarhoven, E. Aarts, Simulated annealing, in *Simulated Annealing: Theory and Applications, Mathematics and Its Applications*, vol. 37 (Springer, Netherlands, 1987), pp. 7–15
15. M. Weiser, Ubiquitous computing. Computer **26**(10), 71–72 (1993)

Automatic Early Risk Detection of Possible Medical Conditions for Usage Within an AMI-System

H. Joe Steinhauer and Jonas Mellin

Abstract Using hyperglycemia as an example, we present how Bayesian networks can be utilized for automatic early detection of a person's possible medical risks based on information provided by unobtrusive sensors in their living environments. The network's outcome can be used as a basis on which an automated AMI-system decides whether to interact with the person, their caregiver, or any other appropriate party. The networks' design is established through expert elicitation and validated using a half-automated validation process that allows the medical expert to specify validation rules. To interpret the networks' results we use an output dictionary which is automatically generated for each individual network and translates the output probability into the different risk classes (e.g., *no risk*, *risk*).

Keywords Ambient assisted living · Bayesian networks · Automated diagnosis

1 Introduction

A major part of the HELICOPTER (Healthy Life support through Comprehensive Tracking of individual and Environmental Behaviors, http://www.helicopter-aal.eu) is to develop information and communication technology (ICT) - based solutions that assist self-sufficient elderly people in early detection of the possible development of medical conditions, such as hyperglycemia or heart failure. The reason for this is to prevent complications arising from the medical conditions if they are not detected early enough. The main contribution of the HELICOPTER project is therefore the part of the system that can detect the risk of certain medical conditions

H. Joe Steinhauer (✉) · J. Mellin
University of Skövde, Skövde, Sweden
e-mail: joe.steinhauer@his.se

J. Mellin
e-mail: jonas.mellin@his.se

A. Mohamed et al. (eds.), *Ambient Intelligence - Software and Applications*,
Advances in Intelligent Systems and Computing 376,
DOI 10.1007/978-3-319-19695-4_2

based on sensor readings and that we call the *automatic triage*. Its system architecture is closer described in [1]. The automatic triage should be as unobtrusive as possible and should not bother the patient with unnecessary interventions. Health surveillance for the automatic triage is achieved by deploying unobtrusive sensors (e.g., infrared sensors, pressure sensors, power meters, body weight scales, and food-inventory tools) and wearable sensors (e.g., fall detectors, individual identification tags). All data collected from these heterogeneous sensors are then interpreted within a data analysis engine in order to deduce the patient's current risk of developing an acute medical condition (e.g., hyperglycemia or hypotension).

In this project it is our objective to utilize well established existing methods, in this case Bayesian networks, deploy them within a case study in order to develop the specific network designs necessary for each medical condition, and validate the resulting networks. The remainder of this paper is organized as follows: In Sect. 2 we explain how a Bayesian network for the use in the automatic triage can be developed in cooperation with a medical expert. After that, in Sect. 3, we describe how the results of Bayesian networks are validated. Last, but not least, we discuss our work and give some suggestions for future work in Sect. 4.

2 Bayesian Networks for Automatic Triage Diagnosis

Generally, a diagnosis will be determined on available evidence E and is defined as in e.g. [2]:

$$d^* = argmax_{d \in D} \Pr(d|E) \quad (1)$$

where D is the set of possible diagnoses, and d^* stands for the subset of diagnoses that have been chosen. Bayesian networks [3] have been used in the area of medical diagnostic reasoning, prognostic reasoning, treatment selection, and for the discovery of functional interactions, since the beginning of 1990 [2, 4, 5]. Some early examples can be found in [4, 6–8]. More recently, Bayesian networks are also applied in home care applications e.g. [9].

A Bayesian network [3] or causal probability network [6] is a graphical representation of a probability distribution over the set of random variables. Probabilistic inference can be done with Bayes rule (see e.g. [10]), which in our domain, where we want to infer the probability of a disease given that we observe one or several symptoms that are often caused by the disease, can be defined as:

$$P(disease|symptom) = \frac{P(symptom|disease)P(disease)}{P(symptom)} \quad (2)$$

Due to their graphical representation, Bayesian networks are relatively easy to understand and to create and can therefore be used, developed, and interpreted by domain experts [9]. They can often be seen as a model of cause-effect relationships [4] whereby their structure and the underlying probability distribution can be learnt from data or be created by hand. Thus qualitative and quantitative knowledge can be mixed [6]. Furthermore, uncertain knowledge can be modeled within a Bayesian network and missing data can be handled during the diagnosis process, which can successively be updated when more evidence becomes available [7].

Before we started to develop the automatic triage system, we also considered alternative evidential frameworks, such as evidence theory [11] and subjective logic [12], but decided together with the medical expert to use Bayesian networks based on four criteria: (1) the framework chosen needs to be able to express everything that is relevant for the task, (2) the design and inner workings of the framework should be easy to understand for the medical expert, (3) the framework should be considerably mature and (4) tools for developing the networks should be available.

In our project, as there is no data set available from that the Bayesian network could be automatically constructed and tested, it needs to be built by hand, whereby knowledge about the domain of diagnosing medical conditions is provided by a medical expert. [2] describes that the construction of a Bayesian network by hand usually involves five stages, which can be iterated during the construction process: (1) relevant variables need to be chosen; (2) relationships among the variables need to be identified; (3) logical and probabilistic constraints need to be identified and incorporated; (4) probability distributions need to be assessed; and (5) sensitivity analysis and evaluation of the network have to be performed.

Expert elicitation is an essential task in order to build the network and goes therefore hand in hand with the network construction. Following [13], expert elicitation is a five step process consisting of: (1) a decision has to be made how information will be used; (2) it has to be determined what information will be elicited from the expert; (3) the elicitation process needs to be designed; (4) the elicitation itself has to be performed; and (5) the elicited information needs to be translated (encoded) into quantities.

A specific problem when working with Bayesian networks is to elicit the prior and conditional probability values. [14] argue that even though probability theory is optimal for the task of decision making, it is often found to be impractical for people to use. On the other hand, qualitative approaches to deal with uncertainty, which appear to be more naturally usable by people, often lack in precision.

In order to elicit the prior and conditional probabilities for our project we developed a dictionary, which, as for example described in [14], can be specified to allow the expert to express his or her belief for or against a statement or claim in a so called argument. The argument is expressed in qualitative terms using qualifiers [14] that then are translated into probabilities. Several dictionaries have been described in the literature (e.g., [15]). However, for our task we needed to develop a suitable dictionary together with the expert, since it was important to the expert to know how the qualitative terms would translate into probabilities in order to fully understand what the qualitative terms stand for. It was also important that the qualitative terms match, as much as possible, the way the expert intuitively thinks about probabilities of symptoms for a developing medical condition. Sometimes we had to reverse the reasoning, since the available information was in the form of $P(symptom|disease)$ rather than $P(disease|symptom)$. The qualifiers and their associated probabilities used are defined as:

- x is known to be false → P(x) = 0
- x is very unlikely → P(x) = 0.01
- x is unlikely → P(x) = 0.1
- x has a negative indication → P(x) = 0.25
- x is random → P(x) = 0.5
- x has a positive indication → P(x) = 0.75
- x is likely → P(x) = 0.9
- x is very likely → P(x) = 0.99
- x is known to be true → P(x) = 1

Note, that this dictionary is only applicable for specifying how probable a medical condition is, given the observable symptoms. To interpret the networks' outcome a different dictionary, which is specific for each individual network needs to be generated. This output dictionary specifies an upper and a lower threshold for the output probability for each risk class (e.g., the classes no risk and risk).

Further information needed from the expert was how the variables depend on each other, what the prior probabilities of the medical conditions and the observables are, what the conditional dependencies between the symptoms and the developing medical condition are, etc. A resulting network for hyperglycemia risk detection for a diabetic person is presented in Fig. 1, developed using GeNIe 2.0 [16]. This network has eight variables in total: food intake increase (FI), body weight gain (BW), soft drink intake increase (SD), gender (G), prostatic hypertrophy (PH), prolapsed bladder (PB), diuresis frequency increase (DF) and risk of hyperglycemia (RH). The latter one is the target variable which probability we are interested in. (A previous and invalidated version of this network can be found in [17].)

The target variable RH provides a probability value which in relation to the aforementioned thresholds indicates if the patient currently is at risk of developing/ experiencing hyperglycemia. The lower threshold for risk of hyperglycemia for this particular network is P(RH) = 0.9 (or true = 90 % for RH) which means that the value of RH for true = 95 %, indicates a risk of hyperglycemia. This result is based solely on the information that is available from the deployed input sources (FI, BW, SD, G, HP, PB, and DF). The network in Fig. 1 shows the case for a diabetic male patient (G male = 100 %) with no prostatic hypertrophy (PH false = 100 %) who has been observed to be drinking a lot of soft drinks (SD true = 100 %). An increased food intake or a body weight gain has not been observed and is therefore set to true = 75 % which is the base rate for either of these that have been derived from the practical experience of the medical expert.

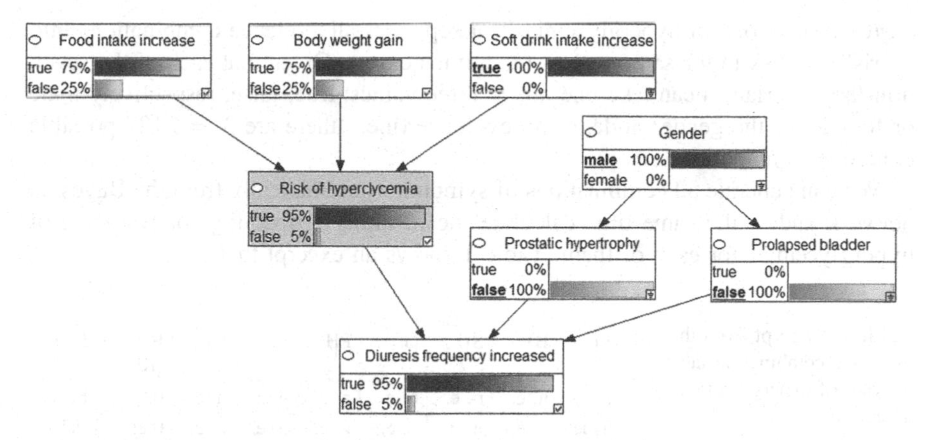

Fig. 1 Bayesian network for risk of hyperglycemia detection

The network in Fig. 1 is only one out of many possible Bayesian networks representing evaluation of risk of hyperglycemia. What variables are part of the network depend on (1) what information that can be provided by sensors and (2) how the experts, involved in the design of the network, perform the diagnosis based on the information from the available sensors. If more or different sensors are used, then the network's layout and the corresponding probability values must be refined. Further, it is important to realize (as previously mentioned) that the network is not performing a complete diagnosis. Instead, it only provides an indication of the risk that the patient is suffering from the effects of hyperglycemia. If the network indicates a risk, the patient will be asked by the AMI system to make sure that they are risking to suffer (or are suffering) from the effects of hyperglycemia by measuring the blood glucose level. The reason to avoid frequent direct measures of the glucose level is to increase the person's quality of life. People quickly tend to become annoyed when they are asked to interact with the system when they can see no obvious reason for it. As a consequence, they may react by generally ignoring the system's recommendations [18]. Therefore, reducing the frequency of and number of interactions whenever possible is important to ensure that the system is used appropriately.

3 Validating the Developed Model

Validity of Bayesian networks established trough expert elicitation is, according to [19], usually tested by comparing the model's predictions to available data or by asking the expert to check whether the network's outcomes appear to be accurate. In our case, we need the expert to specify for each possible combination of evidence, what his diagnosis would be. For the Bayesian network to come to the same result means, that there exists a clear threshold for the probability that the patient currently is at risk of hyperglycemia, that separates all evidence combinations into no risk and risk in the same way as the human expert. Given that each of the seven non-target variables can take one out of three values: true, false, respectively male or female for the gender node, or no evidence (n.e.), there are $3^7 = 2187$ possible evidence combinations.

We can generate all combinations of symptoms automatically from the Bayesian network and at the same time calculate the resulting probability for RH (risk of hyperglycemia) for each of them. Table 1 shows an excerpt thereof.

Table 1 Excerpt from the evidence combination table for risk of hyperglycemia detection

FI	BW	SD	G	PH	PB	DF	P (RH)	Risk
n. e.	n. e.	n. e.	n. e.	n. e.	n. e.	n. e.	0.85	False
True	n. e.	n. e.	n. e.	n. e.	n. e.	n. e.	0.89	False
True	True	n. e.	n. e.	n. e.	n. e.	n. e.	0.92	True
n. e.	n. e.	True	n. e.	n. e.	n. e.	n. e.	0.95	True
n. e.	n. e.	n. e.	n. e.	n. e.	n. e.	True	0.96	True
n. e.	n. e.	n. e.	Male	True	No	True	0.85	False
True	True	True	n. e.	n. e	n. e	n. e	0.99	True
…	…	…	…	…	…	…	…	…

After that, it needs to be identified for which of these cases hyperglycemia actually is suspected, which is represented in the table's last column, denoted Risk. Some of the evidence combinations can be disregarded, as they make no sense. It is, for example, impossible for a patient to have both, prostatic hypertrophy and a prolapsed bladder. For all remaining cases, it needs to be decided if they represent a risk or no risk of hyperglycemia and thereby if the corresponding probability value should be below or above the threshold.

To alleviate this process, validation rules can be specified that cover several alternatives at once and for which the value for Risk then can be set automatically. For example, whenever DF = true and PH = false then Risk = true. This rule covers all cases where the patient suffers from an increase in diuresis frequency but does not have prostatic hypertrophy. Yet another way of formulating rules is to say, e.g. whenever two of the variables FI, BW, and SD are true then Risk = true. These validation rules support the process of partitioning the networks outcome into risk classes. They can be viewed as expressions of criteria for when a risk of the medical condition ought to be detected. These criteria may, however, be incomplete.

The next step is to identify if there is a threshold for P(RH) that clearly partitions all the possible cases into at least two classes one for Risk = true and one for Risk = false. For that, we need to identify the corresponding probability values for P(RH) in each of the partitions. For each partition, we calculate the interval from the lowest value for P(RH) for Risk = false to the highest value for P(RH) for Risk = false and respectively for P(RH) for Risk = true. If the resulting two intervals are non-overlapping, which is the case for the network presented in Fig. 1; we can identify a threshold between these intervals. Each value of P(RH) below this threshold results in no risk of hyperglycemia (Risk = false) and the system not intervening with the person and every value above or equal to the threshold results into risk of hyperglycemia (Risk = true) and the system intervening with the person. If the intervals would overlap, the net is not fully valid to diagnose the risk of the disease without doubt. In that case, the network's design needs to be adjusted accordingly.

Additionally, [19] emphasizes that model validity should not only be checked regarding the model's outcome, but as well regarding the mechanism through that the outcome is obtained. They propose seven different types of validity that the net should be tested for: Nomological validity, face validity, content validity, concurrent validity, convergent validity, discriminant validity, and predictive validity.

As mentioned previously, Bayesian networks have been successfully used within medical diagnostic, which accounts for the nomological validity of our approach. The model's face validity is provided by the expert, who was involved in designing the net, and in analyzing the predictive validity of the net. Content validity is achieved by consulting the expert, rather than the literature. The expert decided what variables and what states of the variables need to be modeled with regard to building a net that models his or her own internal model for risk of hyperglycemia identification. At this stage, the network does not contain any reoccurring parts for that concurrent validity needs to be tested. Convergent and discriminant validity are achieved up to a certain degree through the fact that, as mentioned before, reasoning in medical diagnosis is usually done from symptoms to causes. The world is usually modeled in the way that causes are parent nodes of symptoms. How we achieve predictive validity has been already described above.

4 Discussion and Future Work

In this paper, we described the development of the Bayesian network for automatic detection of a person being at risk of a medical condition on the example of hyperglycemia in a diabetic patient. The purpose of the work presented here is to develop a general method for designing and validating risk detection networks. Deployment of more and different sensors might improve risk detection The network is based on one expert's opinion only and it would therefore be interesting to investigate if a similar network that is based on the elicitation of several experts will show improved results. However, the next step in our project will be to test the network's results against the real world. To identify more risk classes, e.g., no risk, low risk, risk, high risk, very high risk, would be an additional improvement as the system is meant to monitor the patients and to encourage them to a healthier life style. When only a low risk of hyperglycemia is indicated, this could be used to prompt the patient to generally try to change an unhealthy habit that appears to be the reason for the risk being apparent. In order to do that, the system must know what the most likely reason for the diagnosis is. Therefore, explanation methods for Bayesian networks [20] could be applied.

Commonly used methods for information fusion can be roughly grouped into two groups (1) using precise probability e.g. based on Bayesian theory [21] that we have utilized in this approach and (2) using imprecise probability [22] e.g. different variants of evidence theory [11], or credal sets e.g. [23]. These two groups differ from each other regarding how evidence is modeled within the underlying evidential framework and how it is combined [24]. It would be interesting to compare the performance of imprecise frameworks for the same task.

Acknowledgments We would like to thank Giacomo Vespasiani, M.D. for his enthusiastic cooperation. The HELICOPTER project is funded by the Ambient Assisted Living Joint Program (AAL-2012-5-150) (http://www.aal-europe.eu).

References

1. J. Mellin, H.J. Steinhauer, L. Boffi, D. Kristály, P. Ciampolini, N. Pierantozzi, B. Carlsson, C. Berg, M. de Pender, G. Vespasiani, An automated triage for ambient assisted living in HELICOPTER. J. Des. Test (under review)
2. P. Lucas, Bayesian networks in biomedicine and health care. Editor. Artif. Intell. Med. **30**, 201–214 (2004)
3. J. Pearl, Fusion, Propagation and Structuring in Belief Networks. Artif. Intell. **29**, 241–288 (1986)
4. P. Lucas, Bayesian networks in medicine: a model-based approach to medical decision making, in *Proceedings of the EUNITE Workshop on Intelligent Systems in Patient Care* (2001), pp. 73–97
5. S. Visscher, P. Lucas, K. Schurink, M. Bonten, Using a Bayesian-network model for the analysis of clinical time-series data, in *Artificial Intelligence in Medicine 2005*, ed. by S. Miksch et al. LNAI, vol. 358 (Springer, Berlin, 2005), pp. 48–52
6. S. Andreassen, C. Riekehr, B. Kristensen, H.C. Schønheyder, L. Leibovici, Using probabilistic and decision – theoretic methods in treatment and prognosis modelling. Artif. Intell. Med. **15**, 121–134 (1999)
7. D. Aronsky, P.J. Haug, Diagnosing community-acquired pneumonia with a Bayesian network, in *Proceedings of AMIA Symposium* (1998), pp. 632–636
8. I.A. Beinlich, H.J. Suermondt, R.M. Chavex, G.F. Cooper, The ALARM monitoring system: a case study with two probabilistic inference techniques for belief networks, in *Artificial Intelligence in Medicine*, ed. by J. Hunter, J. Cookson J, J. Wyatt. (Springer, 1989), pp. 247–256
9. O. Nee, A. Hein, Clinical decision support with guidelines and Bayesian networks, in *Decision Support Systems Advances*, ed. by G. Devlin. INTECH (2010), pp. 117–136
10. S. Russel, P. Norvig, *Artificial Intelligence: A modern Approach* (Pearson, Upper Saddle River, 2010)
11. G. Shafer, *A Mathematical Theory of Evidence* (Princeton University Press, Princeton, 1976)
12. A. Jøsang, Subjective logic, book draft (2013), http://folk.uio.no/josang/papers/subjective_logic.pdf
13. T.G. Martin, M.A. Burgman, F. Fidler, P.M. Kuhnert, S. Low-Choy, M. Mcbride, K. Mengersen, Eliciting Expert Knowledge in Conservation Science. In: Conservation Biology. Society for Conservation Biology (2011)
14. J. Fox, D. Glasspool, J. Bury, in *Quantitative and Qualitative Approaches to Reasoning Under Uncertainty in Medical Decision Makin.* ed. by S. Quagline, P. Barahona, S. Andreassen. AIME 2001 (Springer, Berlin, 2001), pp. 272–282

15. J. Fox, P. Krause, M. Elvang-Gøransson, Argumentation as a general framework for uncertain reasoning, in *proceedings of the ninth conference on uncertainty in artificial intelligence* (1993)
16. Decision Systems Laboratory, University of Pittsburgh. GeNIe 2.0, http://genie.sis.pitt.edu/?ver=20048430
17. J. Mellin, G. Vespasiani, M. Mustica, G. Matrella, M. Mordonini, C. Berg, I. Schoormans, B. Carlsson, J. Bak, Domain model (2014), http://www.his.se/HELICOPTER-WP4-deliverables/
18. P. Lyons, A.T. Cong, H.J. Steinhauer, S. Marsland, J. Dietrich, H.W. Guesgen, Exploring the responsibilities of single-inhabitant smart homes with use cases. J. Ambient Intell. Smart Environ. **2**(3), 211–232 (2010)
19. J. Pitchford, K. Mengersen, A proposed validation framework for expert elicited Bayesian networks. Expert Syst. Appl. **40**(1), 162–167 (2013)
20. C. Lacave, F.J. Díez, A review of explanation methods for Bayesian networks. J. Knowl. Eng. Rev. **17**, 107–127 (2002)
21. J.M. Bernardo, A.F.M. Smith, *Bayesian Theory* (Wiley, Hoboken, 2000)
22. P. Walley, *Statistical Reasoning with Imprecise Probabilities* (Chapman and Hall, London, 1991)
23. A. Karlsson, R. Johansson, S.F. Andler, Characterization and empirical evaluation of bayesian and credal combination operators. J. Adv. Inf. Fusion **6**(2), 150–166 (2011)
24. A. Karlsson, Evaluating credal set theory as a belief framework in high-level information fusion for automated decision-making. PhD thesis, Örebro University, School of Science and Technology (2010)

Reducing Stress and Fuel Consumption Providing Road Information

Víctor Corcoba Magaña and Mario Muñoz Organero

Abstract In this paper, we propose a solution to reduce the stress level of the driver, minimize fuel consumption and improve safety. The system analyzes the driving and driver workload during the trip. If it discovers an area where the stress increases and the driving style is worse from the point of view of energy efficiency, a photo is taken and is saved along with its location in a shared database. On the other hand, the solution warns the user when is approaching a region where the driving is difficult (high fuel consumption and stress) using the shared database. In this case, the proposal shows on the screen of the mobile device the image captured previously of the area. The aim is that driver knows in advance the driving environment. Therefore, he or she may adjust the vehicle speed and the driver workload decreases. Data Envelopment Analysis is used to estimate the efficiency of driving and driver workload in each area. We employ this method because there is no preconceived form on the data in order to calculate the efficiency and stress level. A validation experiment has been conducted with 6 participants who made 96 driving tests in Spain. The system reduces the slowdowns (38 %), heart rate (4.70 %), and fuel consumption (12.41 %). The proposed solution is implemented on Android mobile devices and does not require the installation of infrastructure on the road. It can be installed on any model of vehicle.

Keywords Intelligent transport system • Fuel consumption optimization • Data envelopment analysis (DEA) • Driving assistant • Android • Android wear • Applications • Mobile computing

V.C. Magaña (✉) · M.M. Organero
Dpto. de Ingeniería Telemática, Universidad Carlos III de Madrid Leganés, Madrid, Spain
e-mail: vcorcoba@it.uc3m.es

A. Mohamed et al. (eds.), *Ambient Intelligence - Software and Applications*,
Advances in Intelligent Systems and Computing 376,
DOI 10.1007/978-3-319-19695-4_3

1 Introduction

Many traffic accidents are due to distractions. In [1] risk factors of traffic accidents are categorized as follows: human factors (92 %), vehicle factors (2.6 %), road/environmental factors (2.6 %), and others (2.8 %). Among these, drivers' human factors consist of cognitive errors (40.6 %), judgment errors (34.1 %), execution errors (10.3 %), and others (15 %).

To reduce traffic accidents due to dangerous behaviors of drivers, it is necessary to investigate, measure, and quantify the drivers' workload. The term "load" in this context indicates the portion of capacity that is needed to drive. This capacity is limited. Therefore, if the task requires a lot of ability is likely that the driver makes mistakes. The level of workload is affected by several factors such as: road type, traffic conditions, driving experience, and gender.

There are many works on measuring and quantifying the driver workload. In [2], Wu and Liu described a queuing network modeling approach to model the subjective mental workload and the multitask performance. They propose to use this model to automatically adapt the interface of driving assistant according to the workload. In [3], Itoh et al. measured electrocardiogram (ECG) signals as well as head rotational angles, pupil diameters, and eye blinking with a faceLAB device installed in a driving simulator to calculate driving workload. In [4], the driver workload from lane changing were measured through simulation test driving. In [5], a multiple linear regression equation to estimate the driving workload was proposed. The model employs variables such as: speed, steering angle, turn signal, and acceleration.

On the other hand, the impact of the cognitive load on the driver behavior has been studied on many papers. In [6], Kim et al. analyzed the relationship between drivers' distraction and the cognitive load. It was discovered that heart rate, skin conductance, and left-pupil size were effective measurement variables for observing a driver's distraction. [7] showed that the visual demand causes a reduction in the speed and increased variation in maintenance lane. However, the cognitive load does not affect speed. In this work the authors highlight that detection of events is very important in order to capture the main safety related effects of cognitive load and visual tasks. [8], the authors propose to use a set of variables (vehicle speed, steering angle, acceleration, and gaze information) to predict the workload driver. The authors achieved an accuracy of 81 % with this method. Other studies [9] propose to use the movement of the steering wheel as an indicator of driver workload.

In conclusion, there are a limited number of works where the workload is analyzed in a real environment driving. Furthermore, there are not applications. The main contribution of this paper is the proposal of an assistant that employs this information about the driver stress and his driving style to build a shared database which contains the places where driving is difficult (high workload and fuel consumption). The objective is to provide knowledge about these places in advance to avoid inefficient actions and improve safety.

2 Discovering Areas Where Driving Required a High Workload

The first step of the proposed algorithm is to find out in which regions the driver is driving inefficiently and stress increases (difficult areas). Data Envelopment Analysis [10] is used to estimate the efficiency of driving and the stress level in each area. Data envelopment analysis (DEA) is a linear programming methodology to estimate the efficiency of multiple decision making units (DMUs) when the production process presents a structure of multiple inputs and outputs. This method was proposed by Charnes, Cooper, and Rhodes [11] (Fig. 1).

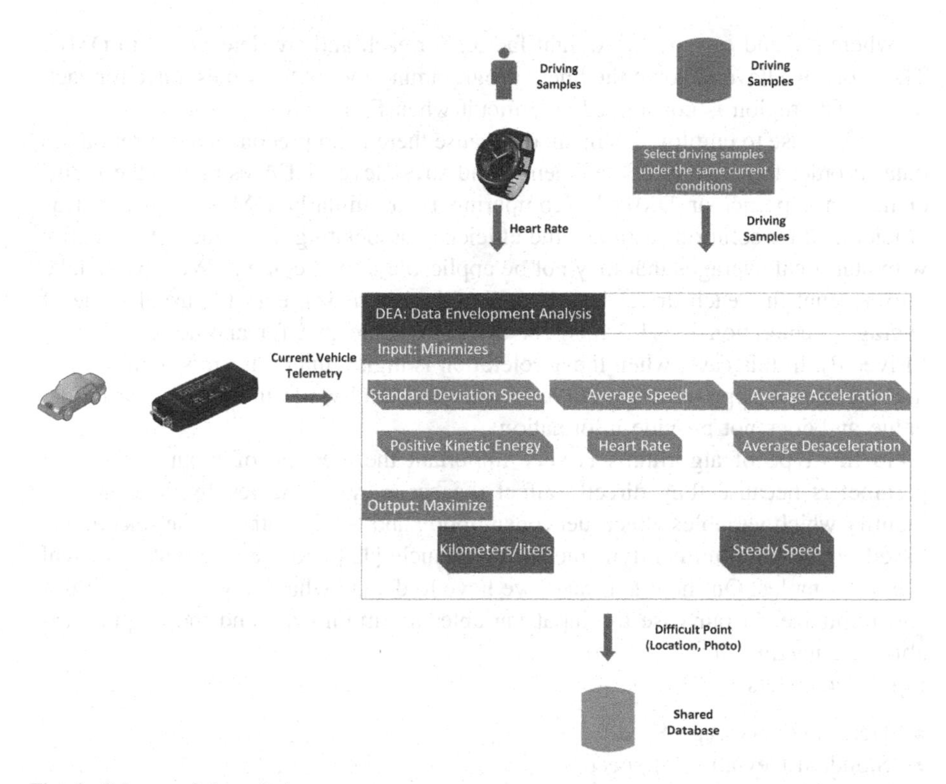

Fig. 1 Schema of the solution

In our proposal, each DMU represents a different driving samples obtained under similar conditions (weather, traffic and road type) by the same driver. The aim is to detect the road points where the driver workload is high and fuel consumption increases. If we consider a set of drivers n (DMU_n), each of them with an I number if inputs and O number of outputs, the efficiency measure E_k for DMU_k is calculated by solving the following linear programming model.

Maximize:

$$E_k = \sum_{o=1}^{O} q_{o,k} \times y_{o,k} \quad (1)$$

Subject to:

$$\sum_{o=1}^{O} q_{o,k} \times y_{o,n} - \sum_{i=1}^{I} p_{i,k} \times x_{i,n} \leq 0; \ \forall n \quad (2)$$

$$\sum_{i=1}^{I} p_{i,k} \times x_{i,k} = 1 \quad (3)$$

$$p_{i,k}, q_{o,k} \geq 0; \forall o, \forall i \quad (4)$$

where $p_{i,k}$ and $q_{o,k}$ are the weight factors for each and are determined to DMU. Therefore, we have to solve the linear programming model "n" times, once for each driver. The region is considered as difficult when E_k is less than 1.

We propose to employ this method because there is no preconceived form on the data in order to calculate the efficiency and stress level. DEA estimates the inefficiency in a particular DMU by comparing it to similarly DMUs considered as efficient. Other solutions estimate the efficiency associating the values of the entity with statistical averages that may not be applicable to that context. We have to take into account that each driver has particular characteristics, e.g.: the usual value of average acceleration is - 1.5 m/s_2 for driver A while that for another is - 1 m/s_2 (driver B). In this case, when the acceleration is higher than - 1.5 m/s_2 could mean that the user is approaching a curve for driver B, and while in case A is a normal value and does not provide information.

In this type of algorithms is very important the election of input and output parameters because they directly affect the accuracy of the results. We have to identify which variables affect fuel consumption and level of stress. The selection is based on the longitudinal dynamics of the vehicle [12] and the observation of real driving samples. On the other hand, we have to decide what we want to maximize and minimize. In our case the input variables are minimized and the output variables are maximized.

Input parameters:

- Heart rate: Average Speed
- Standard Deviation of speed
- Average Deceleration
- PKE (Positive Kinetic Energy):

Output parameters:

- Fuel Consumption (km/l)
- Driving time at steady speed

Positive Kinetic Energy (PKE) measures the aggressiveness of driving and depends on the frequency and intensity of positive accelerations [13]. A low value means that the driver is not stressed and drives smoothly. An unusual high value may indicate that driver are driving in an area that requires special attention such as acceleration lanes or roundabouts. It is calculated using the following equation:

$$PKE = \frac{\sum (v_i - v_{i-1})^2}{d} \quad (5)$$

where v is the vehicle speed (m/s) and d is the trip distance (meters) between v_i and v_{i-1}.

The system takes a photo when it detects that it is difficult driving area where the stress increases and the driving style is less efficient. Photo and location (latitude and longitude) are stored in a shared database. On the other hand, the solution used the shared database in order to warn the driver when he or she is approaching to a difficult area. Therefore, he will know that is coming to a region where it should take precautions and the causes of this warning. The device is fixed on the windshield, where the driver can easily see the screen without taking the eyes off the road. In addition, we may reduce distractions using proposals such as Google Glass or Garmin HUB [14, 15]. These devices allow the user to receive visual information and to pay attention on the road.

3 Evaluation of Proposal

3.1 Experimental Design

The solution was deployed on a LG G3. This device is equipped with a Quad-Core Qualcomm Snapdragon 801 at 2.5 GHz, Bluetooth LE, and 3 GB of RAM. The OBDLink OBD Interface Unit from ScanTool. Net [16] was used to get the relevant data (vehicle speed, fuel consumption, and acceleration) from the internal vehicle's CAN bus [17]. The OBDLink OBD Interface Unit contains the STN1110 chip that provides an acceptable sample frequency for the system. In our tests, we obtain two samples per second. Heart rate was got through LG GWatch-R. This smartwatch run Android Wear and consists of a 1.2 GHz Quad-Core Qualcomm Snapdragon 400 processor, 4 GB internal storage and 512 MB RAM. In addition, it has Bluetooth LE connectivity, barometer, accelerometer, gyroscope, and heart rate monitor (HRM).

In order to evaluate the proposed system, 96 test drives have been performed with 6 different drivers. The tests were performed in Madrid between the months of November 2014 to January 2015. The selected routes (A and B) has both parts of urban road and a highway. All tests were made under similar conditions (time, traffic, and weather). The vehicles employed were all Citroen Xsara Picasso 2.0 HDI.

Drivers were divided into two groups: X and Y. Each group completed a different route (A, B). The experiment consisted of two phases. In the first phase, drivers completed the route 4 times without the use of shared database. At this stage the aim was only to discover areas where stress and fuel consumption is high. Group X drove in route A and group Y drove in route B. In the second stage, the drivers had to drive 4 times in a route different from the first stage. Therefore, group X drove in route B and group Y drove in route A. The objective was that drivers did not know the route. In this case, the solution was activated. The drivers received warnings (photos on the device screen) when they were near a difficult area.

4 Results

Table 1 shows the results obtained in the first phase of the experiment, when the solution was disabled. The objective of this test was only to build the shared database with areas where driving is difficult.

Table 2 captures the results of the second phase. In this case, the system was activated and provided information (photos) to the user when he was approaching a difficult region. As mentioned in the previous section, the drivers drove on different routes from the first phase. Therefore, they did not know the road environment. We can see that the fuel consumption is improved by 12.41 % and the heart ratio is reduced by 4.70 % when the proposal is enabled. In addition, we should highlight that driving is softer (PKE value is lower than in the first phase of the experiment). The reason is that the user can observe the environment in advance and adjust the vehicle speed.

Table 1 Results without using the solution (First Phase)

	Route	Average heart rate (b.p.m)	Std. heart rate (b.p.m)	Fuel consumption (l/100 km)	PKE (m/s2)
Driver X1	A	82.36	8.96	6.93	0.3081
Driver X2	A	75.34	4.20	6.52	0.2995
Driver X3	A	76.10	10.11	6.83	0.3049
Driver Y1	B	76.10	10.07	6.95	0.3052
Driver Y2	B	75.50	4.55	6.51	0.2979
Driver Y3	B	75.13	5.43	6.41	0.2831

Table 2 Results using the solution (Second Phase)

	Route	Average heart rate (b.p.m)	Std. heart rate (b.p.m)	Fuel consumption (l/100 km)	PKE (m/s2)
Driver Y1	A	73.64	3.13	5.85	0.2636
Driver Y2	A	73.60	2.76	5.80	0.2576
Driver Y3	A	73.62	3.89	5.90	0.2599
Driver X1	B	72.60	2.59	5.81	0.2497
Driver X1	B	72.11	2.35	5.86	0.2422
Driver X3	B	73.27	3.03	5.93	0.2562

The major difference introduced by the use of the driving assistant is appreciated in the presence of difficult areas where drive has to adjust the vehicle speed. The results of magnifying the deceleration pattern and heart rate in one of the occasions that the drivers has to slow down is presented in Fig. 2. Graphically, the deceleration rate when using the solution results in a more gradual deceleration pattern. The user has to brake abruptly when he does not receive information about the environment in advance. This causes increased stress and more likely to have a traffic accident. Providing information to the driver is positively correlated with obtaining smooth deceleration patterns in general. The degree of improvement depends on the skill of the driver and his or her response when receiving the warning. Furthermore, smooth deceleration pattern has a positive impact on fuel consumption. In this case, the vehicle takes advantage of the kinetic energy to move until thc point where the vehicle should stop or reduce the speed and is not wasted.

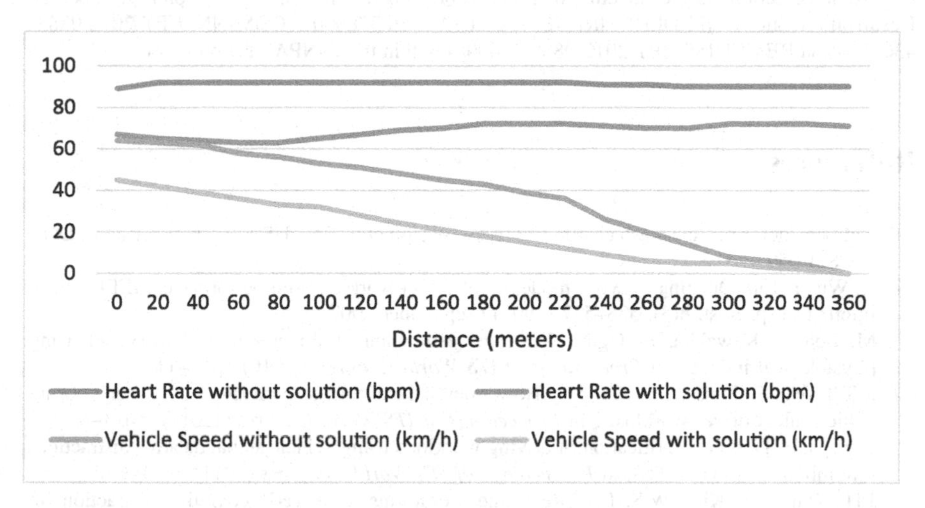

Fig. 2 Deceleration pattern comparison at one difficult area (high fuel consumption and stress) with and without using the solution

5 Conclusions

In this paper, we proposed a method for reducing stress and fuel consumption. The solution builds a shared database with the areas where driving is more difficult based on the driving and the driver workload. Data Envelopment Analysis is used to discover these areas. This method takes into account the particular characteristics of each user. The shared database is used to provide information in advance about the road environment so that driver can adopt appropriate measures. The results show a significant improvement in fuel consumption and a reduction in the driver stress. In the literature, we found a large number of papers about how to measure the driver workload. However, they do not propose any methods to reduce it and tests are conducted in simulators. The main contribution of this work is an application to improve safety and fuel consumption using the information of the level of driver stress and his driving.

As future work, we want to remove the causes that the region is difficult to drive using the captured image, e.g.: detecting that there is a very pronounced curve in the region. This would allow us to issue more precise recommendations and reduce possible distractions caused by the wizard. We also employ other measures to assess the stress of the user as galvanic sensors or camera but always keeping in mind that they should not be intrusive.

Acknowledgments The research leading to these results has received funding from the "HERMES-SMART DRIVER" project TIN2013-46801-C4-2-R within the Spanish "Plan Nacional de I + D+I" under the Spanish Ministerio de Economía y Competitividad and from the Spanish Ministerio de Economía y Competitividad funded projects (co-financed by the Fondo Europeo de Desarrollo Regional (FEDER)) IRENE (PT-2012-1036-370000), COMINN (IPT-2012-0883-430000) and REMEDISS (IPT-2012-0882-430000) within the INNPACTO program.

References

1. National motor vehicle crash causation survey. Washington, DC, USA, Technical Report DOT HS 811 059
2. C. Wu, Y. Liu, Queuing network modeling of driver workload and performance. IEEE Trans. Intell. Transp. Syst. **8**(3), 528–537 (2008) (September 2007.)
3. M. Itoh, E. Kawakita, K. Oguri, Real-time estimation of driver's mental workload using physiological indexes, in *Proceedings of ITS World Congress* (2010) pp. 1–11
4. E.T.T.Teh, S. Jamson, O. Carsten, How does a lane change performed by a neighboring vehicle affect driver workload? in *Proceedings of ITS World Congress* (2012), pp. 1–8
5. S. Sega, H. Iwasaki, Verification of driving workload using vehicle signal data for distraction-minimized systems on ITS, in *Proceedings of ITS World Congress* (2011), pp 1–12
6. J.H. Kim, Y.S. Kim, W.S. Lee, Real-time monitoring of driver's cognitive distraction, in *Proceedings of Spring Conference Korean Society for Automation Engineering* (2011), pp. 1197–1202
7. J. Engström, E. Johansson, J. Östlund, Effects of visual and cognitive load in real and simulated motorway driving. Transp. Res. F, Traffic Psychol. Behav. **8**(2), 97–120 (2005)
8. Y. Zhang, Y. Owechko, J. Zhang, Learning-based driver workload estimation, in *Computational Intelligence in Automotive Applications*, vol. 132, ed. by D. Prokhorov (Springer, Berlin, 2008), pp. 1–24
9. E. Adell, A. Varhelyi, M. Fontana, The effects of a driver assistance system for safe speed and safe distance: a real-life field study. Transp. Res. C, Emerg. Technol. **19**(1), 145–155 (2011)
10. A. Charnes, W.W Cooper, B Golany, L Seiford, J Stutz, Foundations of data envelopment analysis for Pareto-Koopmans efficient empirical production functions. J. Econom. **30**(1–2), 91–107 (1985). ISSN 0304-4076
11. A. Charnes, W.W. Cooper, E. Rhodes, Measuring the efficiency of decision making units. Eur. J. Oper. Res. **2**, 429–444 (1978)
12. I. Ben Dhaou, Fuel estimation model for ECO-driving and ECO-routing, in *Intelligent Vehicles Symposium (IV)* (IEEE, 2011), pp. 37, 42, doi: 10.1109/IVS.2011.5940399. Accessed 5–9 June 2011

13. K.S. Nesamani, K.P. Subramanian, development of a driving cycle for intra-city buses in Chennai, India. Atmos. Environ. **45**(31), 5469–5476 (2011). ISSN 1352-2310, http://dx.doi.org/10.1016/j.atmosenv.2011.06.067
14. Google Glass, http://www.google.com/glass/start/. Accessed 09 Jan 2015
15. H.U.B Garmin, https://buy.garmin.com/en-US/US/prod155059.html. Accessed 09 Jan 2015
16. OBDLink ScanTool, http://www.scantool.net. Accessed 09 Jan 2015
17. S. Godavarty, S. Broyles, M. Parten, Interfacing to the on-board diagnostic system, in *Vehicular Technology Conference, 2000. IEEE-VTS Fall VTC 2000*, 52nd, vol. 4 (2000), pp. 2000–2004. doi:10.1109/VETECF.2000.886162

Policy-Based Adaptation of Context Provisioning in AmI

Alexandru Sorici, Gauthier Picard, Olivier Boissier and Adina Florea

Abstract With the increasing openness and complexity introduced by recent Ambient Intelligence application domains (e.g. Web-of-Things, Sensing-as-a-Service), adaptation of Context Provisioning becomes a key issue. However, methods to easily specify and engineer such mechanisms remain insufficiently explored. In this work we present and evaluate our approach of using semantic-web and multi-agent technologies to define and execute context provisioning policies within our Context Management Middleware called CONSERT.

Keywords Ambient intelligence · Context provisioning · Semantic web · Multi-agent systems · Policy-based control

1 Introduction

Context Management [1] is a well established research area within the Ambient Intelligence (AmI) domain and various proposals of context management middleware (CMM) have been put forward in recent years. However, with increasing openness and complexity expected from AmI applications, the issue of context provisioning control and adaptation remains sparsely explored. Specifically, the way in which context-aware application developers can engineer adaptive provisioning

A. Sorici (✉) · G. Picard · O. Boissier
FAYOL-EMSE, LHC CNRS:UMR5516, Ecole Nationale Supérieure des Mines, 42023 Saint-Etienne, France
e-mail: sorici@emse.fr

A. Sorici · A. Florea
Department of Computer Science, University Politehnica of Bucharest, 313 Splaiul Independentei, 060042 Bucharest, Romania
e-mail: adina.florea@cs.pub.ro

G. Picard
e-mail: picard@emse.fr

O. Boissier
e-mail: boissier@emse.fr

A. Mohamed et al. (eds.), *Ambient Intelligence - Software and Applications*,
Advances in Intelligent Systems and Computing 376,
DOI 10.1007/978-3-319-19695-4_4

mechanisms is insufficiently discussed. To see why context provisioning control is of increasing importance, let us consider the following simple, yet revealing scenario.

The AmI laboratory in room EF210 of university Politehnica of Bucharest can intelligently host different teaching and meeting activities. When empty, only presence sensors are active in the room. Alice walks in, waiting to meet with two other colleagues to discuss the next AmI course. Upon detecting a person, temperature and luminosity sensors are activated, following subscriptions for their updates from AC and projector units. When Bob and Cecille join Alice at the same desk in room EF210, their smartphones deduce they must be in a collective activity and subscribe for such notifications. The AmI Lab manager can deduce ad-hoc meeting situations, but requires noise level and body posture sensors (e.g. Kinect cameras) to be activated. After the meeting is over, the users start leaving the room at different times. As the AmI Lab manager notices the absence of queries for situations that were previously active, it successively starts to cancel inference for ad-hoc meeting and informs sensors in the room that their updates are no longer required, keeping only presence sensors active.

To tackle the control of sensing and inference dynamics as exhibited in the scenario, we propose the use of declarative *context provisioning policies*. We explain our approach based on semantic web and multi-agent technologies to define and flexibly execute these policies. In Sect. 2 we present the foundations of the proposed CMM solution. Provisioning policy definition and execution are detailed in Sects. 3 and 4 and an evaluation of the approach is performed in Sect. 5. We compare with related work in Sect. 6 and conclude the paper in Sect. 7.

2 CONSERT Middleware

The Context Management Middleware CONSERT (CONtext asSERTion) is our solution for offering support for expressive context modeling and reasoning, with flexible deployment and adaptable context provisioning mechanisms. Figure 1

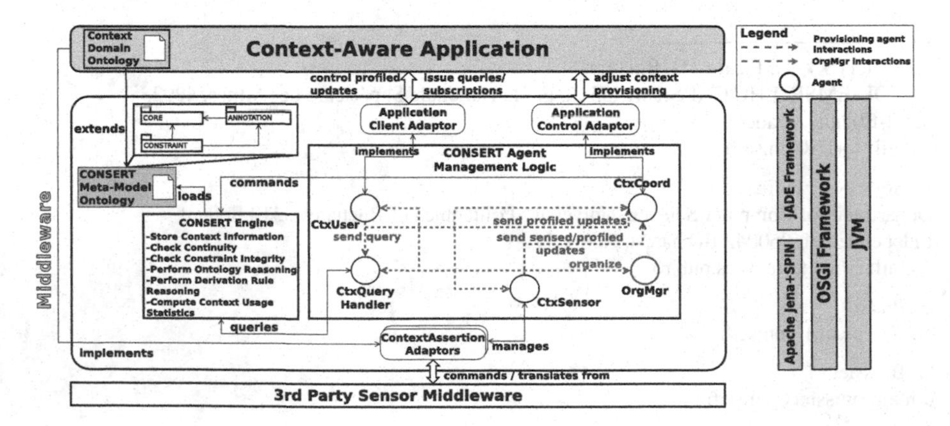

Fig. 1 Conceptual overview of the CONSERT middleware

presents a global view of its architecture and internal management relations. It shows that an ontology-based context model is loaded and used by a reasoning component (CONSERT Engine) which is controlled by an agent-based context management logic. Interaction with application or sensing layers is performed through adaptor services (*ContextAssertion Adaptors* and *ApplicationClient Adaptors*).

Context Modeling and Reasoning We use semantic web technologies as a uniform and expressive mean for context representation and reasoning. Applications build a *context model* using the CONSERT Ontology,[1] which defines a context meta-model [2]. It contains three modules (core, annotation and constraint) able to express context content (i.e. *ContextAssertions* which are the basic model construct describing the situation of *ContextEntities*), context annotations (e.g. source and quality-of-context - QoC metrics) and context integrity, uniqueness or value constraints. For example, *locatedIn(alice, ami-lab){acc:0.8, ts=2015-06-03 12:00:00}* is a *ContextAssertion* specifying Alice's location, where *Person(Alice)* and *University Space(ami-lab)* are *ContextEntities* and *acc:0.8* and *ts=2015-06-03 12:00:00* are *ContextAnnotations* characterizing the accuracy and timestamp of the location statement.

The context model is leveraged at runtime by the CONSERT Engine, which handles context updates, higher-level inference, constraint and consistency checking, as well as asynchronous query answering. The engine inference mechanism performs semantic event processing, employing a rule-based derivation approach using SPARQL CONSTRUCT queries coupled with ontology reasoning. The engine is built as a service component and allows developers to extends its functionality by concrete implementations of constraint resolution and inference scheduling interfaces.

Multi-Agent based Management Architecture Multi-agent design principles allow us to individually encapsulate each aspect of the context provisioning life cycle, thereby opening up the possibility for more flexible management (cf. Sect. 4). Our proposed CMM uses the set of agents shown in Fig. 1. A `CtxSensor` agent interacts with sensors using adaptors to translate from sensor data into statements from the CONSERT Ontology. A `CtxCoord` manages the main life cycle of the CMM using the CONSERT Engine to control context reasoning and consistency. The `CtxQuery Handler` responds to actual dissemination requests, while a `CtxUser` provides the interface with the application specific code above it. Instances of the above agents are deployed within an application to address its context management needs. They form a *Context Management Group* (CMG) and the lifecycle of a CMG is overseen by an `OrgMgr` agent. Next, we show how we define policies that guide and influence the provisioning specific interactions of a CMG (detailed in Sect. 4).

[1] http://purl.org/net/consert-core-ont, http://purl.org/net/consert-annotation-ont, http://purl.org/net/consert-constraint-ont.

3 Context Provisioning Policy

Our approach to context provisioning management is based on *policies* that guide the behavior of the CMG agents. Declarative policy-based control specifications provide best balance between application engineering flexibility and development effort.

In the CONSERT Middleware, semantic web technologies are used again to define context provisioning policies. The CONSERT Provisioning Ontology[2] contains the vocabulary allowing a developer to specify *parameters* and *rules* that govern both the flow of context information within a CMG and the type of inferences carried out by the CONSERT Engine. Designers can specify *what* and *how* information is transmitted and how inference and query handling are prioritized depending on current provisioning requirements (e.g. number of subscriptions, frequency of queries). Provisioning policy vocabulary addresses two processing aspects, sensing and coordination, which we describe next.

3.1 Sensing Policies

Sensing policies define parameters that control *how* updates of sensed information are to be forwarded. Using the CONSERT Provisioning Ontology a developer can specify the *update mode*: *time-based* (send updates at specified time intervals) or *change-based* (send updates on change from a previous value). For *time-based* modes, the update frequency (in seconds) can also be expressed.

After defining an application context model based on the CONSERT Ontology, the developer will assign management of sensed *ContextAssertions* to a set of `CtxSensor` agents within a CMG. These agents use adaptor services to communicate with a physical sensor in custom ways. However, it is the sensing policies that instruct how captured sensor readings are further sent to the `CtxCoord` agent. The excerpt below shows an example of a presence sensing configuration from our reference scenario, specifying a time-based update every 2 s. It governs the update mode for a *ContextAssertion* detecting device bluetooth addresses.

```
:presenceSensingPolicy
  a sensorconf:SensingPolicy ;
  coordconf:forContextAssertion  smartclassroom:sensesBluetoothAddress ;
  sensorconf:hasUpdateMode coordconf:time-based ;
  sensorconf:hasUpdateRate 2 .
```

3.2 Coordination Policies

Sensing policies specify the desired default context update modes. However, the dynamics of the provisioning process, as seen in the reference scenario, may require an adaptation of how and what information is exchanged. *Provisioning coordination policies* define *settings* and specify *actions* that address such dynamics. A provisioning coordination policy has two components: *control parameters* and *control rules*.

[2]http://purl.org/net/consert-provisioning-ont/control, http://purl.org/net/consert-provisioning-ont/sensing.

Control parameters govern relevant settings of the context provisioning process. They affect both context information transmission and inference processes. In terms of transmission, control parameters specify *which ContextAssertions* currently need to be enabled and how long a particular instance of a *ContextAssertion* must be kept in the CONSERT Engine working memory (its TTL). With respect to inference, the parameters configure the currently active derivation rules (by specifying enabled *derived ContextAssertions*), the type of inference scheduling service (e.g. first-come first-served - FCFS, priority based), the type of constraint resolution service (e.g. prefer-newest, prefer-accurate) or the interval at which to execute ontology-based reasoning. Within a coordination policy, these parameters can have both a general as well as a *ContextAssertion*-specific form overriding the general one. The excerpt below, taken from our reference scenario, shows an example control parameter specification.

```
:ProvisionPolicy_EF210 a coordconf:ControlPolicy ;
  coordconf:hasInferenceSchedulingType coordconf:FCFS ;
  coordconf:hasSpecificAssertionEnabling
    [ a coordconf:AssertionSpecificEnableSpec ;
      coordconf:forContextAssertion person:locatedIn ;
      coordconf:hasParameterValue true
    ] ;
```

Apart from transmission and inference, the *observation_window* parameter configures the length of the time window over which statistics of context information and inference usage are computed by the CONSERT Engine. This parameter can again have both general and *ContextAssertion*-specific configurations. The statistics gathered by the CONSERT Engine together with snapshots of its current knowledge base constitute the triggering conditions of the *provisioning control rules* which can alter the value of control parameters.

Control rules are the concrete means to specify *adaptation actions* of the context provisioning process. They are given as SPARQL CONSTRUCT queries which construct instances of *OperationalCommands*. These commands can have a general effect (e.g. alter the type of inference scheduling service) or a *ContextAssertion*-specific one (e.g. start/stop updates, alter update mode). In the evaluation section we show an example of a control rule template and how instances of control rules are assigned in a coordination policy. The control rules developer can use CONSERT Engine supplied statistics to compose the body of the rules. During each observation window period, the engine keeps track of elements such as: number of total and successfully executed queries and inferences, number of active subscriptions and time elapsed since last query per *ContextAssertion*.

Control rules may have contradictory outcomes (e.g. one rule implies enabling a derivation while another one disables it). To help developers keep a consistent result, the control rules can be partitioned into ordered execution groups. This is achieved using a feature of the SPIN[3] specification, that allows declaring an order relation over ontology properties. Thus, the rules from later execution groups override contradictory results from rules in earlier groups.

[3]http://spinrdf.org/.

Note that currently the CONSERT Provisioning Ontology can express control parameters and rules at a level of *ContextAssertion* types. This means that an *OperationalCommand* demanding the activation of updates for the *hasNoiseLevel ContextAssertion* will target all `CtxSensor` agents that provide such updates, regardless of their potential quality. In future work we aim to provide the ability for *OperationalCommands* to target individual context providers and context derivation rule instances based on quality of context aspects.

4 Context Provisioning

Provisioning Interactions The context provisioning process is composed of two main interaction chains. The sensing chain concerns the updates that `CtxSensor` agents send to the `CtxCoord` of a CMG. The request chain regards queries and subscriptions that `CtxUsers` make to the `CtxQueryHandler` of a CMG. These two chains interact with each other through the working of the CONSERT Engine. As mentioned in Sect. 2, the `OrgMgr` deploys and oversees the lifecycle of a CMG. Once started, they communicate with the `OrgMgr` using standard FIPA[4] Requests to determine their partners (`CtxSensors` with `CtxCoord` and `CtxUsers` with `CtxQueryHandler`) for a bilateral interaction governed by a specific protocol.

In the *sensing chain*, when `CtxSensor` agents start, they register with the `CtxCoord` and look for a sensing-specific policy file to figure out *how* they must provide their updates. They then start a FIPA Propose protocol to publish their *ContextAssertion* update capabilities to the `CtxCoord`. The `CtxCoord` acknowledges the publishing and returns the subset of *ContextAssertions* for which the sensor agent currently has to send updates, according to the control parameters that the coordinator reads from the provisioning coordination policy. Updates are then sent using FIPA Inform messages.

Within the *request chain*, a `CtxUser` that registered with a `CtxQueryHandler` can ask for information via FIPA Query or FIPA Subscribe protocols. However, the response on the query handler side depends on whether the requested *ContextAssertions* are currently enabled or not (according to the provisioning coordination policy). If enabled the `CtxQueryHandler` poses the query against the CONSERT Engine and returns the response. Otherwise, the protocol continues with the query handler notifying the `CtxCoord` of an activation need. In the case of a sensed *ContextAssertion* the coordinator determines which *CtxSensors* can provide it and uses a *TaskingCommand* request to tell them to start sending updates. In case of derived context, it uses the CONSERT Engine command interface to enable the corresponding derivation rules. Only when the `CtxQueryHandler` has the confirmation of *ContextAssertion* enabling will it pose the query to the CONSERT Engine.

Provisioning Control Execution At runtime, the `CtxCoord` agent executes the indexed provisioning control rules at intervals depending on the general *observation_window* parameter. Execution occurs as explained in the previous section and

[4] http://www.fipa.org/.

the CtxCoord collects the output of the *OperationalCommands*. If these demand information flow changes (e.g. start *ContextAssertions* updates, alter update frequency), the CtxCoord determines all *CtxSensors* that provide the concerned assertions and sends them a *TaskingCommand* request wrapping over the content of the *OperationalCommand*. If the operational commands imply changes in inference handling (e.g. enable a derived assertion, change derivation priority service) the CONSERT Engine command interface is used to perform the required adaptation.

5 Scenario Evaluation

Let us now analyze the effect of context provisioning management on sensing behavior and the CONSERT Engine reasoning cycle in the context of the scenario from the introduction. We use iCasa[5], an execution platform on top of OSGi for digital home applications, to model the scenario environment. We simulate the layout of the EF210 smart room and the functionality of 29 sensors (presence, noise level, body posture, temperature, luminosity). We also use iCasa to script the execution of the scenario (succession of events). However, the coordination of the context provisioning process is done using a real deployment of CONSERT Middleware agents, working as explained in Sects. 2 and 4.

Provisioning policy setup Two configurations are used to evaluate the impact of using context provisioning policies. The first policy enables all *ContextAssertion* updates by default, without using *ContextAssertion* monitorization. The second uses provisioning control parameters to enable only presence sensing and location derivation by default. All other context information updates are activated and deactivated on demand. Notifications from the CtxQueryHandler control update activation, while four provisioning control rules from the policy set the conditions for disabling specific updates and derivations.

Figure 2 shows an excerpt in Turtle[6] syntax from the file defining the provisioning control parameters and control rules, used by the CtxCoord agent of the EF210 room.

```
CONSTRUCT {
  _:b0 a :StopRuleCommand.
  _:b0 :forContextAssertion ?assertion.
}
WHERE {
  ?stat a :AssertionSpecificStatistic.
  ?stat :forContextAssertion ?assertion.
  ?stat :isDerivedAssertion true.
  ?stat :nrSubscriptions 0.
  ?stat :timeSinceLastQuery ?time.
  FILTER (?time > ?elapsedThreshold).
}
```

```
coord:ControlPolicy
  coord:hasStopAssertionCommand [
    a  :QueryAbsenceDerivationCancellation;
    arg:contextAssertion ami:AdHocDiscussion;
    arg:elapsedTimeThreshold 10;
  ];
  coord:hasStopAssertionCommand [
    a :NonePresentAssertionCancellation;
    arg:elapsedTimeThreshold 40;
  ];
```

Fig. 2 SPARQL expression of derivation cancellation rule template (left) and control rule assignment (right)

[5] http://adeleresearchgroup.github.io/iCasa/.

[6] http://www.w3.org/TeamSubmission/turtle/.

On the left of the figure we see the SPARQL expression of a rule cancellation template that says that rules that derive a *ContextAssertion* with no current subscriptions and no queries received since *?time* seconds ago will be cancelled. The right side of the figure shows how instances of such templates (e.g. for *ami:AdHocDiscussion*) are assigned in a provisioning coordination policy. The *NonePresentAssertionCancellation* control rule specifies that if no person has been located in the room for more than 40 s, all *ContextAssertion* updates will be ceased.

Provisioning policy influence Figure 3 shows the influence of provisioning policies on sensing behavior and update messages. In the simulation, the sensing policy configures updates every 2 s. As expected, the case where all sensor updates are enabled by default (left hand graph) shows an almost constant sensing activity (2563 sensor readings and 1168 message updates during the 3 min simulation). The right hand graph (corresponding to adaptive provisioning) shows a dynamic evolution of the sensing behavior. At first, only presence sensors are active. When the first person enters the room, luminosity and temperature sensors become active (interval 1 on chart), following queries from the projector and air conditioning units. When all 3 persons are at a desk talking, they subscribe for possible activity notifications and the noise level and body posture sensors needed for ad-hoc meeting detections become active (interval 2). When the meeting is over, the provisioning control rules ensure that unnecessary sensing activity be ceased (interval 3), leaving only presence sensing enabled. Overall, this reduces the sensing activity for the same simulation by 30%. In a real environment, this could have significant benefits in terms of network load and power consumption.

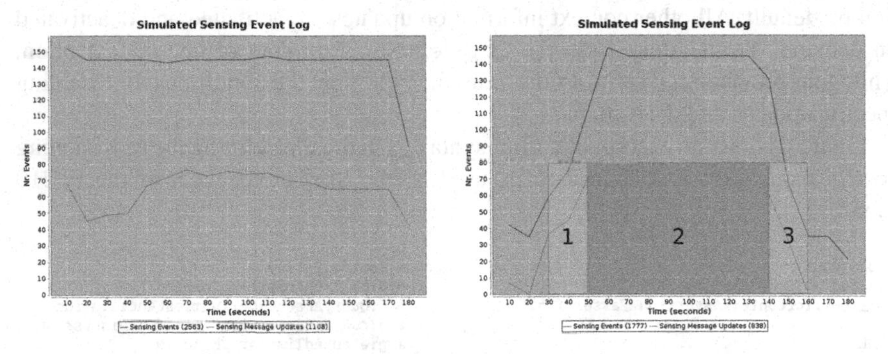

Fig. 3 Sensing events and update messages graph. No provisioning control (left), with provisioning control (right)

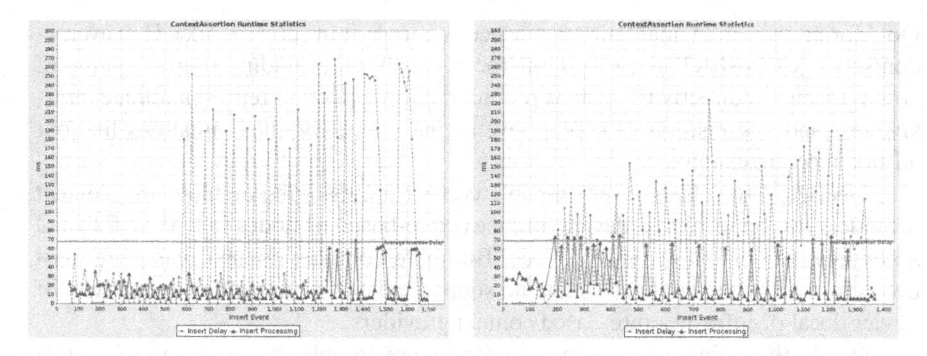

Fig. 4 CONSERT engine update performance showing insertion delay (red) and insertion processing (blue) times. No control rules (left), with provisioning control (right)

Figure 4 shows measurements of CONSERT Engine insertion delay (time until processing starts) and processing duration for *ContextAssertion* updates. While the average delay and processing times for the two simulation configurations are very similar, on the right hand graph we observe slightly higher insertion processing times (blue) and a greater density of insertion delay points (red) that are higher than the average. This is a consequence of having an additional periodic context knowledge base transaction for the evaluation of provisioning control rules. However, these results stem from the particular setup of our simulation (relatively reduced load and information diversity) and a technical limitation of the current implementation which relies on making transaction snapshots of the entire knowledge base when executing insertions, inferences or provisioning control rules. Furthermore, if the dynamic between active and non-active update and derivation periods is greater than the number of control rules, we expect that insertion delay and processing times will actually improve.

6 Related Works

Many CMM solutions have been proposed in recent years. However, the problem of supporting an adaptive context provisioning process remains insufficiently addressed. COSINE [3] is a service-oriented framework for context-awareness in mobile environments. It is therefore mostly addressed to mobile ad-hoc network (MANET) scenarios and strives to provide adaptive query routing based on QoC aspects. However, the rules by which the routing is done are predetermined and cannot be influenced by the application developer as in our case. SOLAR [4] and SALES [5] are two context management solutions which incorporate policy-driven context provisioning adaptation. Both approaches allow applications to monitor and specify policies that control system resource usage (SALES) and operator functionality (SOLAR) in order to adapt context data distribution. These works consider

QoC for provisioning adaptation (an aspect we look at in future work). However, in contrast to CONSERT, these systems focus mostly on providing proper routing of context requests directly to the best provider, thus neglecting representation expressiveness and aggregation of context events into more complex situations like our ad-hoc meeting example.

CONTORY [6] offers a very flexible context provisioning architecture (context providers can be local to a device, infrastructure-based or nodes in a MANET) and an expressive context access interface. But more complex context reasoning must be performed by the application and no support exists for controlling the activity of device-local or infrastructure-based context providers.

In works that, similar to our own, perform more complex aggregations of context situations, context provisioning control is not fully explored. COPAL [7] introduces a domain specific language for context processor deployment and operation specification. However, provisioning control is limited to filtering and aggregation functions which only affect the type of data received by consumers. As opposed to our approach, no context usage statistics are available to processors, so that they may alter the behavior of active providers. Finally, CoCA [8] is a CMM solution with context modeling and reasoning capabilities similar to our own. CoCA uses heuristic methods to ensure loading of only relevant context data into an ontology-based reasoner. However, the heuristics cannot be controlled and, contrary to CONSERT, the middleware offers no mean to adapt information flow.

7 Conclusions

In this work, we propose an approach for adaptation of context provisioning in AmI applications. We propose both Context Provisioning Policies and Context Provisioning Interaction Protocols that govern the interactions and processing of provisioning agents in a CMG. These capabilities increase application-development support by moving context provisioning adaptation concerns away from application and towards the middleware level. Presented examples relate to the dynamics of active and non-active context updates and derivations, but the possibilities of our approach far exceed these cases. The multi-agent based architecture of our context provisioning units implies individual encapsulation of provisioning concerns, holding the potential for increased autonomy. In future work we wish to explore the concept of context-level agreements in the context of provisioning adaptation. This means augmenting our provisioning policies with the ability to express conditions and actions affecting *individual* context providers and inference rules based on observed QoC and application-specified operation goals for each context provisioning agent.

References

1. M. Knappmeyer, S.L. Kiani, E.S. Reetz, N. Baker, R. Tönjes, Survey of context provisioning middleware. IEEE Commun. Surv. Tutor. **15**(3), 1492–1519 (2013)
2. A. Sorici, O. Boissier, G. Picard, A. Zimmermann, *Applying Semantic Web Technologies to Context Modeling in Ambient Intelligence*. Evolving Ambient Intelligence. (Springer International Publishing, Dublin, 2013)
3. L. Juszczyk, H. Psaier, A. Manzoor, S. Dustdar. Adaptive query routing on distributed context-the cosine framework, in *10th International Conference on Mobile Data Management: Systems, Services and Middleware* (2009), pp. 588–593
4. G. Chen, M. Li, D. Kotz, Data-centric middleware for context-aware pervasive computing. Pervasive Mob. Comput. **4**(2), 216–253 (2008)
5. A. Corradi, M. Fanelli, L. Foschini, Adaptive context data distribution with guaranteed quality for mobile environments, in *5th IEEE International Symposium on Wireless Pervasive Computing (ISWPC)* (2010), pp. 373–380
6. O. Riva, Contory: a middleware for the provisioning of context information on smart phones, in *Proceedings of the ACM/IFIP/USENIX 2006 International Conference on Middleware* (2006), pp. 219–239
7. F. Li, S. Sehic, S. Dustdar, Copal: an adaptive approach to context provisioning, in *Wireless and Mobile Computing, Networking and Communications (WiMob)* (IEEE, 2010), pp. 286–293
8. D. Ejigu, M. Scuturici, L. Brunie, Hybrid approach to collaborative context-aware service platform for pervasive computing. J. Comput. **3**(1), 40–50 (2008)

An Adaptive Particle Filter for Indoor Robot Localization

Hao Lang, Tiancheng Li, Gabriel Villarrubia, Shudong Sun and Javier Bajo

Abstract This paper develops an adaptive particle filter for indoor mobile robot localization, in which two different resampling operations are implemented to adjust the number of particles for fast and reliable computation. Since the weight updating is usually much more computationally intensive than the prediction, the first resampling-procedure so-called partial resampling is adopted before the prediction step, which duplicates the large weighted particles while reserves the rest obtaining better estimation accuracy and robustness. The second resampling, adopted before the updating step, decreases the number of particles through particle merging to save updating computation. In addition to speeding up the filter, sample degeneracy and sample impoverishment are counteracted. Simulations on a typical 1D model and for mobile robot localization are presented to demonstrate the validity of our approach.

Keywords Particle filter • Monte Carlo localization • Mobile robot

H. Lang · T. Li · S. Sun
School of Mechanical Engineering, Northwestern Polytechnical University, Xi'an 710072, China
e-mail: langhao@mail.nwpu.edu.cn

S. Sun
e-mail: sdsun@nwpu.edu.cn

T. Li (✉) · G. Villarrubia
BISITE Group, Faulty of Science, University of Salamanca, Salamanca 37008, Spain
e-mail: t.c.li@mail.nwpu.edu.cn

G. Villarrubia
e-mail: gvg@usal.es

J. Bajo
Department of Artificial Intelligence, Technical University of Madrid, 28660 Madrid, Spain
e-mail: jbajo@fi.upm.es

A. Mohamed et al. (eds.), *Ambient Intelligence - Software and Applications*, Advances in Intelligent Systems and Computing 376,
DOI 10.1007/978-3-319-19695-4_5

1 Introduction

The particle filter (PF), utilizing sequential Monte Carlo (SMC) approach to implement the Bayes estimation, has been widely applied for nonlinear systems such as target tracking and mobile robot localization since [1], which is able to relax the linearity and Gaussian assumptions. However, the PF suffers from high computational burden when a large number of particles are required. This paper is concerned with designing an adaptive PF that can adaptively adjust the number of particles for efficient computation while maintains good estimation performance. The proposed PF is applied for robot localization in an indoor structural environments with the use of a group of sonars and an inertial measurement unit (IMU)-based odometer.

When applied to robot localization, the PF is often referred to as the Monte Carlo localization (MCL) method. It represents the required robot position and orientation by a set of random particles with associated weights. Let $S_t = \{x_t^i, w_t^i\}_{i=1,2,\ldots N_t}$ denote a random measure that characterizes the posterior density $p(x_t|y_{1:t})$:

$$p(x_t|y_{1:t}) \approx \sum\nolimits_{i=1}^{N_t} w_t^i \delta(x_t - x_t^i), s.t. \sum\nolimits_{i=1}^{N_t} w_t^i = 1 \quad (1)$$

where δ is Dirac delta measure, $\{x_t^i\}_{i=1,2,\ldots N_t}$ is a set of particles with associated weights $\{w_t^i\}_{i=1,2,\ldots N_t}$, N_t is the total number of particles.

The weights w_t are determined with respect to the principle of sequential importance sampling (SIS), which relies on

$$w_t \propto \frac{p(x_{1:t}|y_{1:t})}{q(x_{1:t}|y_{1:t})} \approx w_{t-1} \frac{p(y_t|x_t)q(x_t|x_{t-1})}{q(x_t|x_{t-1}, y_t)} \quad (2)$$

where $q(\cdot)$ is a proposal importance density, which should resemble the posterior density as closely as possible.

Usually, the weight variance will exponentially increase with time in the SIS, causing weight degeneracy that a few particles have very large weights while the others are negligible. As such, the resampling step is often required which reset the particle system in order to solve the degeneracy. But one critical side effect may arise in the meanwhile, namely sample impoverishment, see [2, 3] i.e. most particles are of the same state(s) that are duplicated from a few particles while the other particles of small weight are abandoned in the resampling process. This has much the same effect as sample degeneracy, and is more severe when the measurement noise is small. Many works have been devoted to solve this pair of problems, see [3].

For mobile robot localization, the weight updating of particles is much more time-consuming than the prediction, which is the main reason for the huge computation requirement of the PF. Thus, the key to speed up the PF is to reduce the number of particles for updating. This, however, conflicts with that a large number of particles are required in the prediction step for accurate approximation. To alleviate this contradiction, we propose a double resampling strategy to adjust the number of particles for prediction and updating: use a large number of particles for prediction while use a small number of particles for updating. In brief,

- To obtain a compromise of the contradiction between the computing speed and the estimation accuracy, we use different number of particles at the prediction and updating step respectively through double resampling methods.
- To alleviate sample degeneracy and impoverishment, the resampling should maintain the diversity of particles.

The work presented here is a significant extension of the work [4]: we replace the first traditional resampling used therein with the partial resampling to avoid discarding any particle, in order to maintain better the diversity of particles. In addition, we improve the second particle merging resampling by proposing a novel grid dividing method for better computational speed. The paper is organized as follows. Related work is given in Sect. 2. Two resampling procedures are detailed in Sect. 3. Simulation results are presented in Sect. 4 before we conclude in Sect. 5.

2 Related Work

A heavy computation requirement will cause that "the rate of incoming sensor data is higher than the update rate of the PF and so some sensor data will be missed" [5]. Most solutions to improve the speed of the PF are focused on adjusting the number of particles to reduce the number of iterations required. There are various same size adaptation mechanisms, see the survey [6]. One of the most elegant methods for adjusting the number of particles is KLD-sampling approach [7, 8] and KLD-resampling [9]. Furthermore, Legland [10] proposed to adjust the number of particles by ensuring that a sufficient number of samples whose weights are large enough are used by the filter. It prevents that particles are located in regions of the state space having zero posterior probability. Pan [11] used rate-distortion theory to determine the optimal particle number, Fitzgerald [12] proposed an advanced proposal scheme by reducing the number of particles needed for multiple target tracking.

To adjust number of particles has to be careful. When the PF is running with a small sample set, it becomes challenging to approximate the posterior distribution properly. For this purpose, the characters of particle are pre-studied and stored, which will accelerate the updating of particles greatly in [13]. However, these pre-stored particles have introduced errors. An up-to-date review of the parallel resampling methods and parallel PFs is available in [2].

There is another idea [14] that uses clusters of particles to track multiple distinct hypotheses, where each cluster is considered as an independent hypothesis about the robot's pose. The algorithm works on two different levels: at the particle level, the classical Bayesian formulation is adopted to update a hypothesis, while at the cluster level, the one with the highest probability is used to determine the robot's pose.

A good idea is to use different number of particles at different steps of prediction and updating respectively. Following this line of thinking, we reduce the number of particles for updating but increase the number of particles of prediction through two times of resampling while maximally alleviating the degeneracy and impoverishment.

3 Double Resampling Strategies

The iteration of the general PF can be divided into three steps: prediction, updating and resampling. In contrast, the proposed PF in this paper consists of four steps: 1st resampling, prediction, 2nd resampling and updating. The pseudo-code of the proposed adaptive PF can be described as algorithm 1. The double resampling structure aims to maintain the sample quality (avoiding sample degeneracy and impoverishment) while adjust the number of particles. In specific,

- The 1st resampling, partial resampling (Step 2 of Algorithm 1), is executed before the prediction step, which duplicates particles with large weight (larger than a threshold) while reserves particles with small weight (instead of discarding). Here, each particle of weight $w_t^{(i)}$ is duplicated with $N^{(i)}$ times,

$$N^{(i)} = \left\lfloor N \times w_t^{(i)} \right\rfloor \tag{3}$$

where $\lfloor \cdot \rfloor$ means rounding down operation, N is the desired sample size (a reference). It can be seen only the particle of weight that is smaller than $1/N$ will be duplicated otherwise will be reserved.

Since there is no particle discarded, it is easy to know that the distribution does not change at all and the diversity of the particle population will be well maintained. It will however increase the sample size. However, this will not increase computing time heavily, since the computation of the prediction step is only a small part in the MCL. For this, we employ the second dynamic grid resampling to control the sample size.

- The 2nd resampling conducts particle merging before the updating step to reduce the same size to save computation. The updating step is the most time-consuming part of the PF, so particle merging can save weight updating time by decreasing the number of particles. The particle merging is proofed to be unbiased [15] and enjoys good identical distribution attribute, which will not undermine much the particle diversity.

In the following, we shortly explain how to perform the particle merging based on new dynamic grid dividing of the state space with regard to MCL. Denote the planar position and orientation of the robot in the state space as $\{x, y, \theta\}$ and the estimate $\hat{x}_t^i = (\hat{x}^i, \hat{y}^i, \hat{\theta}^i)^T$. The dynamic grid partitioning is realized in three steps: (1), sort the particles according to their states on each dimension; (2) group the particles into different grids each of which containing specified n particles; the grids $g_{x,y,\theta}$ are of differently sizes; (3) merge particles using (4) in each grid as did in deterministic resampling [15], where $(\hat{x}_t^p, \hat{w}_t^p)$ denotes the support particle, $p = p_{x,y,\theta}$ denotes the number of particles in grid $g_{x,y,\theta}$. The pseudo-code of the merging resampling can be described as in Algorithm 2. The partitioning of grids is much computationally easier than [15]. The parameter n can be the same, or different for different dimensions. For simplicity, we use the constant parameter n in different dimensions.

Algorithm 1: The proposed adaptive particle filter for MCL

Input:
- $S_{t-1} = \{(x_{t-1}^i, w_{t-1}^i)\}_{i=1}^{N_{t-1}}$, particle set
- u_{t-1}, control measurement
- y_t, observation

Output:
- New particle set at time t: $\{(x_t^i, w_t^i)\}_{i=1}^{N_t}$

Procedure:

1. 1st Resampling (Partial resampling)
   - $J = 0;$
   - for $i = 1:\hat{N}_{t-1}$
     - $N^{(i)} = floor(N_{t-1} \times w_{t-1}^i) + 1$
     - for $i = 1:N^{(i)}$
       - $J = J + 1$
       - $\hat{x}_{t-1}^J = x_{t-1}^i$
       - $\hat{w}_{t-1}^J = w_{t-1}^i / N^{(i)}$
     - end for
   - end for
2. Prediction
   - for $i = 1:J$
     - $(\hat{x}_t^i, \hat{w}_t^i) \sim p(x_t | x_{t-1}, u_{t-1})$
   - end for
3. 2rd Resampling
   - $\{(x_t^i, \tilde{w}_t^i)\}_{i=1}^{N_t} = (\{(\hat{x}_t^i, \hat{w}_t^i)\}_{i=1}^J, i)$
   - (DG-resampling see Algorithm 2)
4. Updating
   - for $i = 1:N_t$
     - $w_t^i = p(y_t | x_t^i) * \tilde{w}_t^i$
   - end for
5. Normalize the weights
   - for $i = 1:N_t$
     - $w_t^i = w_t^i \left[\sum w_t^i\right]^{-1}$
   - end for.

Algorithm 2: Dynamic Grid Resampling (DG-resampling)

Input:
- $\hat{S}_{t-1} = \{(\hat{x}_t^i, \hat{w}_{t-1}^i)\}_{i=1}^{\hat{N}_{t-1}}$, particle set after RR
- n, the number of grids to use

Output:
- $\tilde{S}_t = \{(\tilde{x}_t^i, \tilde{w}_t^i)\}_{i=1}^{N_t}$

Procedure:
- $p_{x,y,\theta} = zeros\left(3, \left\lceil \frac{\hat{N}_{t-1}}{i} \right\rceil\right)$
- $\{g_{x,y,\theta}\} = \{\}$
- $\tilde{S}_t = \{\}$
- where $\lceil \cdot \rceil$ means rounding up operation.

Procedure

1. Sort operations of particles in different dimensions:
   - $[X^i, \sim] = sort(\hat{x}^i)$
   - $[Y^i, \sim] = sort(\hat{y}^i)$
   - $[\Theta^i, \sim] = sort(\hat{\theta}^i)$
2. Dividing variable space grids:
   - for $i = 1 \rightarrow \hat{N}_{t-1}$ do
     - if $(X^{1+(kx-1)n} \le \hat{x}^i \le X^{kx*n}$ $\& Y^{1+(ky-1)n} \le \hat{y}^i \le Y^{ky*n}$ $\& \Theta^{1+(k\theta-1)n} \le \hat{\theta}^i \le \Theta^{k\theta*n})$, do
       - $p_{kx,ky,k\theta} = p_{kx,ky,k\theta} + 1$
       - $g_{kx,ky,k\theta} = g_{kx,ky,k\theta} \cup (\hat{x}_t^i, \hat{w}_{t-1}^i)$:
       - $(\hat{x}_t^i, \hat{w}_{t-1}^i) \rightarrow (x_t^{p_{kx,ky,k\theta}}, w_t^{p_{kx,ky,k\theta}})$
     - end if
   - end for
   - where $kx, ky, k\theta \in N^+ \& kx, ky, k\theta \le \left\lceil \frac{\hat{N}_{t-1}}{n} \right\rceil$
   - if $kx * n > \hat{N}_{t-1}$ do $X^{kx*n} = X^{\hat{N}_{t-1}}$, end if
   - if $ky * n > \hat{N}_{t-1}$ do $Y^{ky*n} = Y^{\hat{N}_{t-1}}$, end if
   - if $k\theta * n > \hat{N}_{t-1}$ do $\Theta^{k\theta*n} = \Theta^{\hat{N}_{t-1}}$, end if
3. Merge particles in non-empty grid: ($p \triangleq p_{kx,ky,k\theta}$)

$$\tilde{S}_t = \{(\tilde{x}_t^i, \tilde{w}_t^i)\}_{i=1}^{N_t} : \begin{cases} \tilde{x}_t^p = \sum_{p=1}^{P} x_t^p w_t^p / \tilde{w}_t^p \\ \tilde{w}_t^p = \sum_{p=1}^{P} w_t^p \end{cases} \quad (4)$$

where N_t is the total number of non-empty grids.

4 Simulations

4.1 1D Simulation Model

For the sake of evaluating the adaption ability of the number of particles that the proposed double resampling will provide for the PF, we study the following non-linear time-varying framework with the state transition function as

$$x_t = 0.5x_{t-1} + \frac{25x_{t-1}}{1 + x_{t-1}^2} + 8\cos(1.2(t-1)) + e_t \tag{4}$$

and the observation function as

$$y_t = 0.05x_t^2 + v_t \tag{5}$$

where e_t and v_t are zero-mean Gaussian noises with variances 10 and 1 respectively. We use root mean square error (RMSE) to evaluate the estimate accuracy. The simulation length is 1000 steps in each trial.

In our instance, several typical resampling methods/strategies are compared with the present SIRR approach. Sorted by the times of resampling at each iteration, they are basic selective resampling PF (resample or not), SIR PF (once) and SIRR PF (twice) in addition to the known KLD resampling and deterministic resampling. In the following, we will firstly give a brief introduction of their parameter setting.

The selective resampling strategy resamples only when the variance of the non-normalized weights is superior to a pre-specified threshold namely the Effective Sample Size (ESS) criterion and the resampling is implemented only when it is below a threshold N_T, typically $N_T = N_t/2$ in our instance.

In the KLD resampling [9], the KLD-bound parameters $\varepsilon = 0.10, \delta = 0.01$ and the grid size is $L_{kld} = 0.5$ in our case, N_{KLD} is further hard-limited to be no more than $2N_{st}$. For the deterministic resampling [15], parameters are set as: $L_{star} = 1$, $L_{min} = 0.1$, $a = 8$, and $N_{min} = \min(50, N/5)$.

In the simulation, all filters use the same starting number of particles, i.e. their numbers of particles associated are initially the same. For different starting number of particles from 20 to 200, the number of particles of the selective resampling and the SIR does not change over time but they vary in the KLD resampling, Deterministic resampling and our SIRR approach. In particular, the number of particles is constant in the iteration in all the given PFs except in the SIRR approach the number of particles is different at the prediction stage and the updating stage. This is the unique performance of our approach different to others. This is shown in Figs. 1 and 2, where the number of particles of the SIRR has been plotted separately for the prediction and the updating. It shows that the average number of particles in the prediction stage is significantly larger than at the updating stage in the SIRR while they are the same for all the other PFs. These indicate the SIRR manages to work as supposed. It is necessary to note that, the number of particles of the deterministic resampling is slight smaller than the reference sample due to the particle merging, while the number of particles obtained in SIRR is significantly reduced. The number of particles of the KLD-resampling maintains at the stable level. They are based on different mechanism to adjust the number of particles and the SIRR leave much more space for the filter designer.

The RMSE of all filters are given in Fig. 3. Resampling is critical for this model as shown that the selective resampling obtains obvious lower accuracy than always resampling. The estimate accuracy of different resampling strategies/methods are very similar for this model. But the RMSE of the KLD resampling is slight higher when the starting same size is large since its number of particles is the smallest (maintained at the same level). Since all these resampling methods are unbiased and they do not change the particle population much, this simple model is computationally simple and therefore has not shown the advantage of our approach. The SIRR aims to improve the computing efficiency by reducing the times of updating computation, as it only works when the resampling time is much less than the weight updating computation of particles. As stated, the updating step of MCL is much more time-consuming than the simple prediction step, then our method will work. This will be illustrated in our following simulation where the weight updating of particle is computationally more intensive than the prediction and resampling.

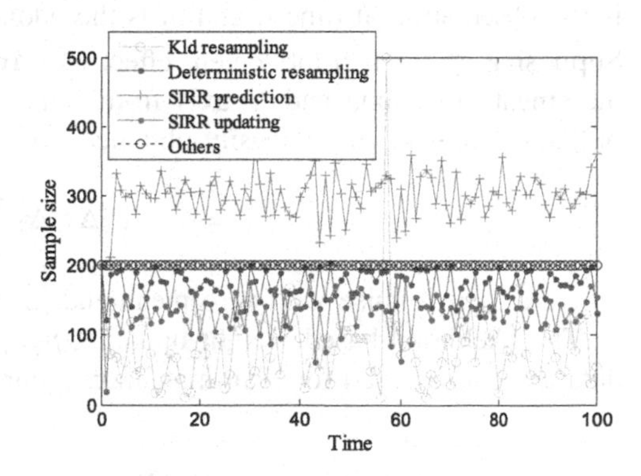

Fig. 1 The number of particles over time in different PFs

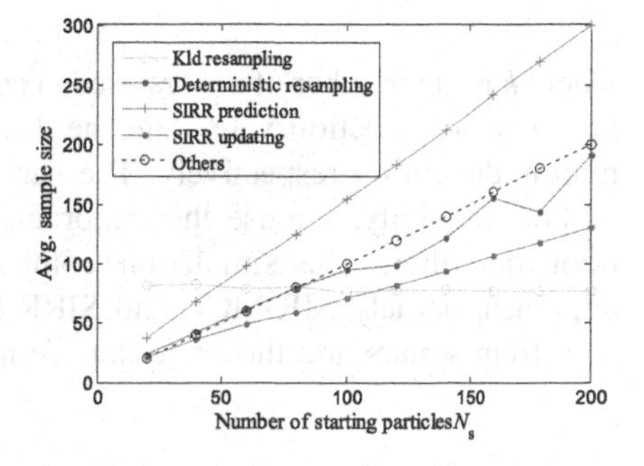

Fig. 2 The average number of particles

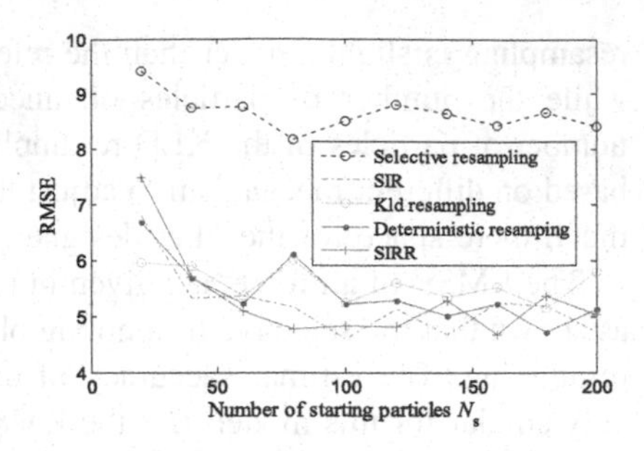

Fig. 3 The average RMSE

4.2 Mobile Robot Localization Simulation

In this section, we will realize the proposed adaptive PF for indoor mobile robot localization simulation, in order to evaluate its computational efficiency and accuracy. The simulation model is showed in Fig. 4. In the figure, the black area indicates the area occupied by obstructs. Most areas of the environment are well modelled and static, while there are still disturbances from human being and some un-modelled areas (represented by the grey circle and rectangular).

Let $x_t = (x, y)^T$ be the robot's position in Cartesian space (x, y) at time instant t, y_t is the observation at time t, and u_t is the odometer data between time $t-1$ and t. Supposing u_{t-1} has a movement effect $(\Delta x, \Delta y)^T$ on the robot, Δx and Δy are the increment of X axis and Y axis from time $t-1$ to t. Then, the motion model $p(x_t|x_t - 1, u_t - 1)$ can be easily obtained from

$$x_t = x_{t-1} + [\Delta x, \Delta y]^T + e_{t-1} \tag{6}$$

where e_{t-1} is a noise with zero mean and $[\Delta x \times 10\,\%, \Delta y \times 10\,\%]^T$ variance.

The likelihood-based weight updating $p(y_t|x_t)$ based on the nearest-neighbour data association used for scan matching in our case can be described as

$$p(y_t|x_t) \propto \prod_{i=1}^{k} \exp\left(- \frac{(z_i - h_i)^2}{2 \times r_i^2} \right) \tag{7}$$

where k is the number of sonars used. The robot uses 8 sonars for observation. z_i and h_i are sonar detection value and the distance between particles and boundaries of map in direction i respectively. The variance of sonar detection value $r_i = z_i * 5\,\%$.

For simplicity, we use the importance density $p(x_k|x_{k-1})$ as proposal, namely bootstrap filter. The simulation compares the basic SIR filter and our SIRR approach, namely SIR-MCL and SIRR-MCL. They receive the same observation data from sonars and the same data from the odometer at the same time.

In the simulation, the robot is supposed to follow a planned path as given in Fig. 4. This planned path consists of two straight lines and a part of a circle. Because of disturbances and process noises, the real path of robots are not really the planned path as supposed. Both MCL algorithms use 1000 as initial number of particles, and set the parameter $n = 30$ in the second merging-resampling of SIRR-MCL. They receive 70 times of observations in the whole process, obtaining different localization paths. To capture the average performance, 10 times of trials are performed based on the same planned path. The results are given in Fig. 5 for average paths of both MCL algorithms. Their estimate error as compared with true path are given in Fig. 6. The average processing time of them is given in Fig. 7.

The results show that the average processing time of SIRR as less as one third of that of the SIR filter while maintain similar accuracy because the SIRR efficiently reduce the number of particles for updating while keep adequate particles for prediction. The results show that the introduction of merging error before updating is beneficial for the computational efficiency and our approach provides a choice for particular cases where computing speed is more important.

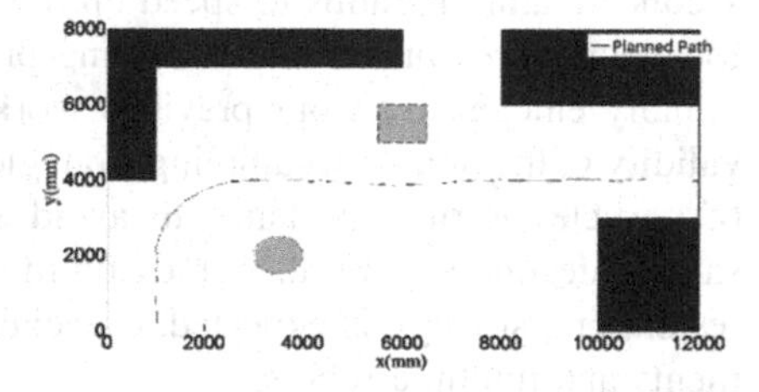

Fig. 4 Indoor environment and robot moving path

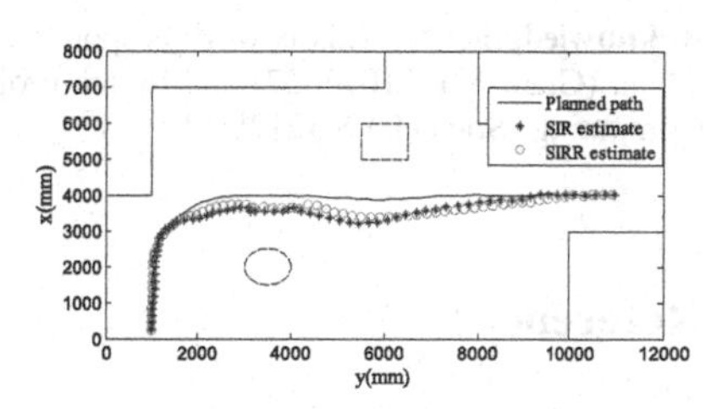

Fig. 5 Localization results of path

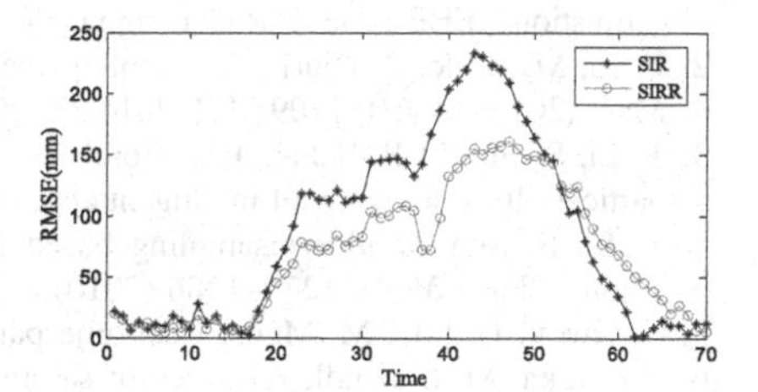

Fig. 6 Localization errors

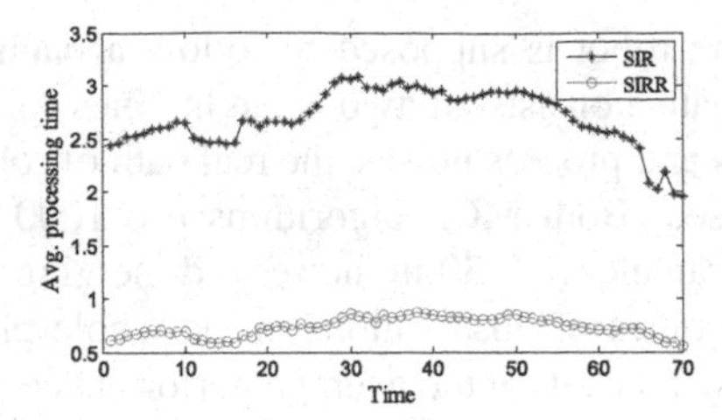

Fig. 7 Average processing time

5 Conclusion

An adaptive PF is proposed for indoor mobile robot localization, which employs resampling two times to overcome the contradiction between computational speed and filtering accuracy. The first resampling is aimed at improving the accuracy of the PF as a large number of particles are used for more accurate prediction, and the second resampling aims to speed up the filter by reducing the number of particles for weight updating. Both resampling procedures are new and are more computationally efficient than our previous work [4]. Simulations have demonstrated the validity of the double-resampling strategies. We reiterate that the spatial distribution of particles is of importance to avoid sample impoverishment while combating sample degeneracy. Further study on more efficient implementations of the double resampling strategy is expected, especially regarding to unstructured real environments and multiple robots.

Acknowledgments This is work is sponsored partly by National Natural Science Foundation of China (Grant No. 51075337) and by the project Sociedades Humano-Agente en entornos Cloud Computing (Soha+C) SA213U13.

References

1. N. Gordon, D. Salmond, A. Smith, Novel approach to nonlinear/non-Gaussian Bayesian state estimation. IEEE Proc. F Radar Signal Process. **140**(2), 107–113 (1993)
2. T. Li, M. Bolic, P. Djuric, Resampling methods for particle filtering. IEEE Signal Proccess. Mag. (2015). doi:10.1109/MSP.2014.2330626
3. T. Li, S. Sun, T.P. Sattar, J.M. Corchado, Fight sample degeneracy and impoverishment in particle filters: a review of intelligent approaches. Expert Syst. Appl. **41**(8), 3944–3954 (2014)
4. T. Li, S. Sun, Double-resampling based Monte Carlo localization for mobile robot. Acta autom. Sinica **36**(9), 1279–1286 (2010)
5. C. Kwok, D. Fox, M. Meilă, Real-time particle filters. Proc. IEEE **92**(3), 469–484 (2004)
6. O. Straka, M. Simandl, A survey of sample size adaptation techniques for particle filters, in *15th IFAC Symposium on System Identification*, vol. 15, Part 1 (2009)
7. D. Fox, Adapting the sample size in particle filters through KLD-sampling. Int. J. Robot. Res. **22**(12), 985–1003 (2003)

8. A. Soto, Self adaptive particle filter, in *Proceedings of International Joint Conferences on Artificial Intelligence* (2005), pp. 1398–1406
9. T. Li, S. Sun, T.P. Sattar, Adapting sample size in particle filters through KLD-resampling. Electron. Lett. **46**(12), 740–742 (2013)
10. F. Legland, N. A Oudjane, Sequential algorithm that keeps the particle system alive. Technical report, Rapport de recherché 5826, INRIA (2006)
11. P. Pan, D. Schonfeld, Dynamic proposal variance and optimal particle allocation in particle filtering for video tracking. IEEE Trans. Circuits Syst. Video Technol. **18**(9), 1268–1279 (2008)
12. M. Orton, W. Fitzgerald, A Bayesian approach to tracking multiple targets using sensor arrays and particle filters. IEEE Trans. Signal Process. **50**(2), 216–223 (2002)
13. T. Li, S. Sun, Y. Gao, Localization of mobile robot using discrete space particle filter. J. Mech. Eng. **46**(19), 38–43 (2010)
14. A. Milstein, J. Sánchez, E. Williamson, Robust global localization using clustered particles filtering, in *Proceedings of the 18th National Conference on Artificial Intelligence*, (Edmonton, Alberta, Canada, 2002)
15. T. Li, T.P. Sattar, S. Sun, Deterministic resampling: unbiased sampling to avoid sample impoverishment in particle filters. Sig. Process. **92**(7), 1637–1645 (2012)

A Discomfort-Sensitive Chair for Pointing Out Mental Fatigue

André Pimenta, Davide Carneiro, Paulo Novais and José Neves

Abstract In our busy daily life, we often have the feeling of being exhausted, accompanied with a sense of performance degradation and increase of discomfort in the execution of even simple tasks. This often takes place in the workplace and in a silent way, influencing our productivity, our performance the number of errors or the quality of our production. This paper details a chair to be used in workplace environments that is sensitive to the onset of fatigue. Based on built-in accelerometers it recognizes signs of discomfort, which may be related to mental fatigue, to point out moments when an individual should consider taking a pause or a rest. This chair complements a previously developed software for the assessment of mental fatigue from the analysis of the individual's interaction with the computer.

Keywords Ambient intelligence · Fatigue · Statistical analysis · Clustering

1 Introduction

Until recently, most of the jobs available were essentially physical. However, this paradigm has changed and, in just a few decades, a significant part of this offer became mental and cognitive. Consequently, mental fatigue has replaced physical fatigue, becoming one of the major health-related issues nowadays. This issue is further exacerbated by current increased competition, precarious work conditions and information overload, which add to the existing stress and pressure in the workplace [1].

A. Pimenta (✉) · D. Carneiro · P. Novais · J. Neves
ALGORITMI-Universidade do Minho, Braga, Portugal
e-mail: apimenta@di.uminho.pt

D. Carneiro
e-mail: dcarneiro@di.uminho.pt

P. Novais
e-mail: pjon@di.uminho.pt

J. Neves
e-mail: jneves@di.uminho.pt

A. Mohamed et al. (eds.), *Ambient Intelligence - Software and Applications*,
Advances in Intelligent Systems and Computing 376,
DOI 10.1007/978-3-319-19695-4_6

In addition to these factors, many of these jobs are the so-called desk-jobs, in which people frequently sit for more than 8 hours. This is due to the increasing job specialization which requires the worker to perform only one function or movement for a long period of time, or due to the nature of the work to be performed, that requires the interaction with the computer [2, 3].

These current work conditions often result in constant feelings of tiredness or exhaustion, going beyond what would be considered a normal degree of sleepiness or discomfort. People will often ignore or even try to hide such these feelings, regarding as normal signs of our lifestyle. However, their persistence for prolonged periods may be a sign that something more serious is wrong [4].

The increase of this type of feelings and the negative consequences that they have both on the individual and on the organization led to the development of systems that increasingly care about the user, and its context in order to improve the work environment [5]. An example is Fatigue Audit InterDyne (FAID®) [6], that generates mental performance predictions based on work hours, rest periods and circadian cycles. The CogSpeed system[1] (Gray Matter Metrics, San Antonio, Texas), on the other hand, provides an estimate of the highest speed with which the subject can respond to a discrete stimulus, as a way of tracking critical tasks. Another example of this current trend is the system NovaScan™, that is a performance-based testing procedure designed to provide industry with a brief screen for detecting potential employee job-related performance decrement, including drug use, alcohol use, or fatigue [7].

This paper details a chair developed with the aim of supporting an existing fatigue management software [8]. While the software is based on the analysis of the interaction patterns of working individuals with the computer to assess fatigue, the chair will use accelerometers to identify signs of discomfort from the movements of the worker, signs which are generally associated with the onset of fatigue. The ultimate goal of this line of research is to develop more positive work environments, both for the worker and for the organization.

2 Case Study

A case study was developed to test the hypothesis that the movements of a worker, while sitting in the chair, can be correlated with the level of mental fatigue felt. The participants, nine men in total, were students and researchers from the University of Minho who volunteered to participate. Their age ranged between 18 and 30. The only requirements asked of the participants were that they would use the accelerometer-equipped office wheeled chairs while carrying out their work at the computer. Other than that they were only requested to perform their activities as usual, without any restriction whatsoever.

[1]The website of CogSpeed is available at https://www.cogspeed.com/.

2.1 Methodology

Two accelerometers were placed in the office chairs, as detailed in Fig. 1, to record the movements of the workers during the day (8 hour workday), while sitting in front of the computer. Accelerometer 1 was placed at the level of the worker's back, with the aim to register the acceleration caused by movements of the torso. Accelerometer 2 was placed in one of the wheeled arms, with the aim to record acceleration generated by the moving of the chair. Specifically, Axivity's WAX3 wireless accelerometers were used.[2]

Simultaneously with the monitoring of the chair movements, workers answered a questionnaire about mental fatigue on a hourly basis (USAFSAM Fatigue Scale [9]). This was implemented with the aim of studying, in parallel, the daily evolution of mental fatigue and the movements of the chairs (Fig. 2).

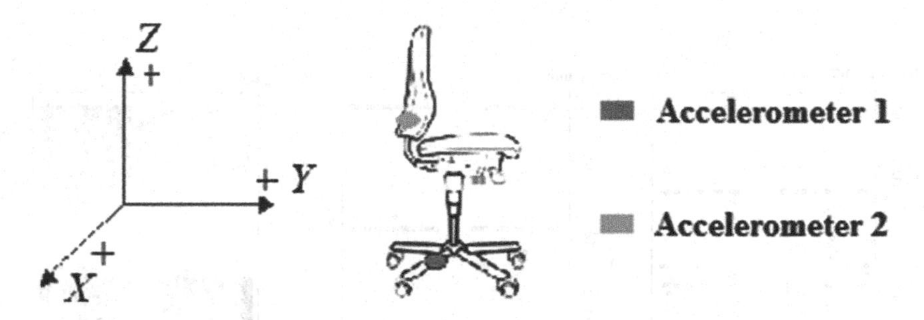

Fig. 1 The placement of the accelerometers in the chairs and the directions of the axes

2.2 Data Analysis

The first step in the process of data analysis was to determine if there was indeed an increase in the level of fatigue during the day. As it can be seen in Fig. 3, the average value of mental fatigue pointed out by the workers increases throughout the day. Interestingly enough, it can be also be observed that the lunch break caused a temporary decrease of the subjective feeling of fatigue.

Having determined this, we proceeded by splitting the data from the accelerometers in hourly series, so as to match the intervals at which questionnaires were collected. Using the Anderson-Darling test, it was determined that the distributions of the eight resulting datasets were not normal. Given this, the Kruskal Wallis test was used in the subsequent analysis. This test was used to compare the distributions of each of the three axes and of the two accelerometers. The maximum p-value observed was 2.2e-16, which demonstrates that the changes in acceleration that occur from one interval to the next are statistically significant.

[2]The website of Axivity is available at http://axivity.com/v2/.

The difference in the distributions can also be visually observed as in Fig. 4. This figure depicts violin plots, which in addition to the information provided by box-plots include the probability density for the different values. These plots show the evolution of the distributions of the values during the day.

The Fig. 2 offer a different perspective on the data, which hints at increased movement as the day goes by, providing the first assumption that when fatigued, workers tend to move more, on a sign of physical and mental discomfort. This is confirmed by the values detailed in Table 1.

Table 1 Sizes of the dataset generated from all the nine workers

	9H	10H	11H	12H	13H	14H	15H	16H	17H
Accelerometer 1	11546	3464	4402	X	11674	3903	6578	6581	6841
Accelerometer 2	11828	4436	6130	X	14197	4797	7205	8520	10435

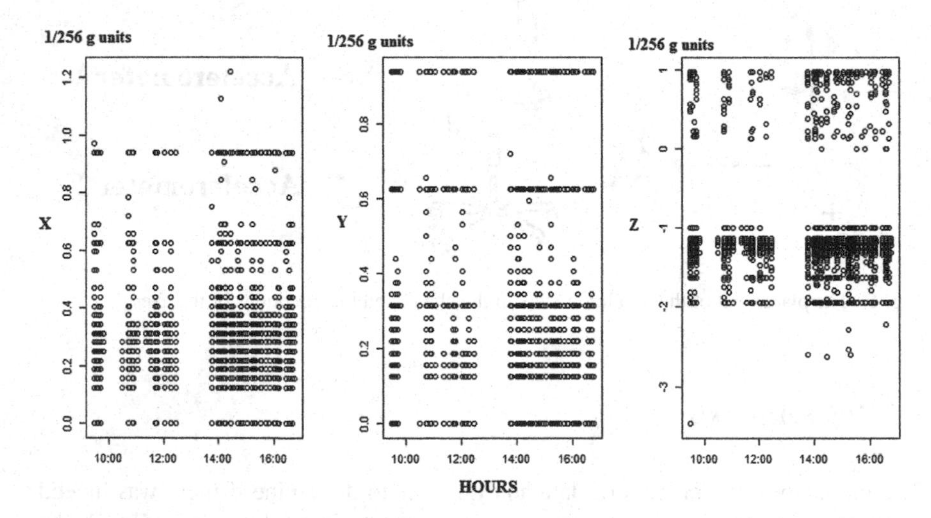

Fig. 2 Distribution of the values of accelerometer 2

Having carried out this preliminary analysis of the data, it was examined whether there is any relationship between thc subjective feeling of mental fatigue and the acceleration measured on the chair, in both positions. To this end it was used the Pearson's test to determine the statistical correlation between the acceleration measured in each axis and the subjective level of fatigue. The results obtained are shown in Table 2.

It can be concluded that the readings of accelerometer 1 have a strong positive correlation with the level of mental fatigue for the axes X and Y. Concerning accelerometer 2, a strong negative correlation with the subjective feeling of fatigue exists for the y-axis.

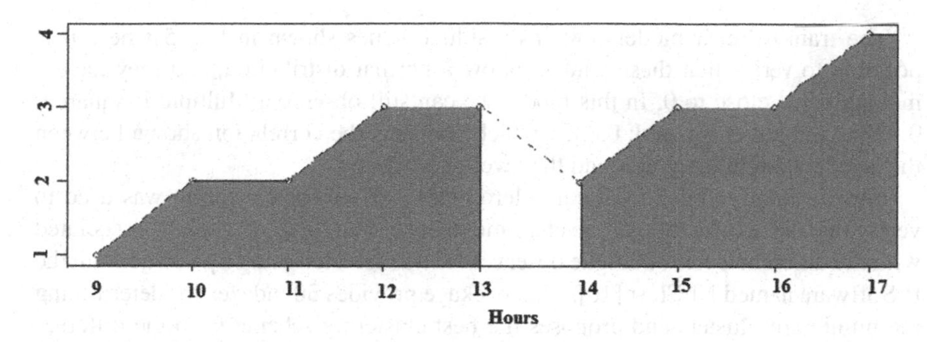

Fig. 3 Average level of fatigue recorded throughout the day by using a questionnaire with the USAFSAM scale

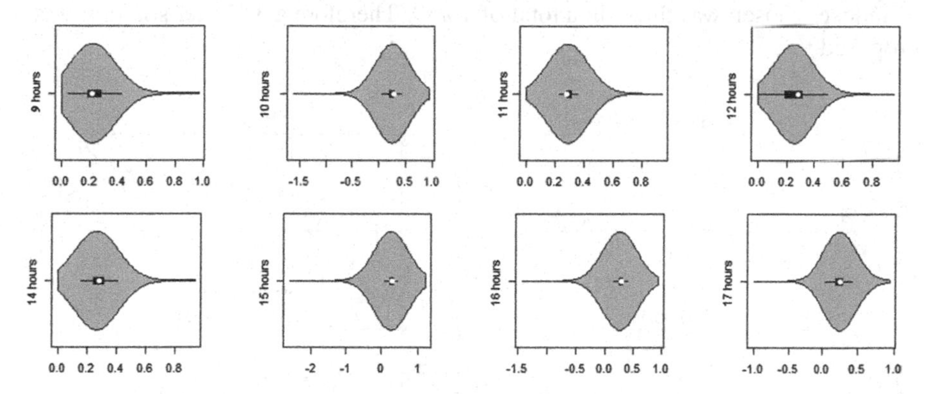

Fig. 4 Hourly violin plot of x-axis for accelerometer 2

Table 2 Correlation between each axis and the subjective level of fatigue, using the pearson correlation coefficient

	x-axis	y-axis	z-axis
Accelerometer 1	0.67	0.53	−0.25
Accelerometer 2	0.194	−0.9	−0.11

2.3 Results

In the preliminary analysis of the data it was established a relationship between the subjective levels of mental fatigue and the movement of the chair in some of its axes in both accelerometers. For modelling accelerometer 1, which showed a strong negative correlation with fatigue for axes X and Y and no significant correlation for the z-axis, a linear regression was fit to the average acceleration values on these axes. While the z-axis is left out due to a lack of correlation, its absence is also positive in the sense that acceleration from when the user sits down or stands up (which happens mostly on this axis and is not the normal behaviour we are observing) will not influence the model.

The trained linear model contains residual values shown in Fig. 5 where it is possible to verify that these values follow a normal distribution, and they have a median value close to 0. In this model we can still observe a Multiple R-squared: 0.8385, Ajusted R-squared: 0.8331 which confirms the correlation shown between the acceleration in the y-axis and the level of fatigue.

For the data recorded by the accelerometer 2, K-means clustering was used to verify that the acceleration of the chair movement in all three axes can be associated with the different levels of fatigue observed. K was selected using a package from the R Software named NbClust [10]. This package provides 30 indexes for determining the number of clusters and proposes the best clustering scheme from the different results obtained by varying all combinations of number of clusters, distance measures, and clustering methods. The suggested value of K with the largest number of indexes chosen was three, in a total of 10/30. Therefore a 3-cluster solution was proposed.

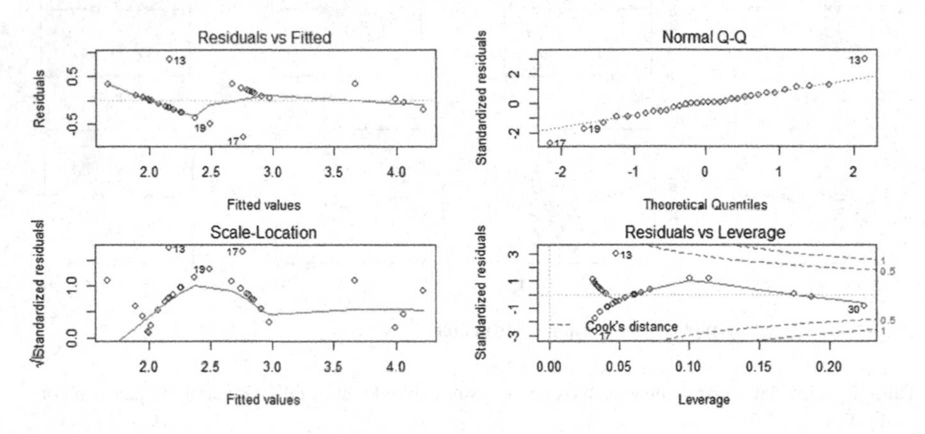

Fig. 5 Linear model built for accelerometer 2

With a defined K the final solution was obtained, as depicted in Fig. 6 and Table 3. The accuracy of the resulting clusters was computed through the ajusted Rand index. The adjusted Rand index provides a measure of the agreement between two partitions, adjusted for chance. It ranges from −1 (no agreement) to 1 (perfect agreement). Agreement between data labelled and the cluster solution is 0.56.

Table 3 Centroid values of the clusters

Cluster	x-axis	y-axis	z-axis
1	0.2671764	0.30934943	−1.149943
2	0.2628069	0.68517492	−1.177082
3	0.2442235	0.02766418	−1.208630

The results obtained for accelerometer 2 show that it is possible to separate the collected data into three distinct clusters. While workers pointed out four different levels of fatigue, the first level was only chosen 10 % of the times. This division thus make sense also when considering the answers to the questionnaires.

3 Conclusions

This paper presented an analysis of the movement of nine workers on their office chairs in the workplace, during their workday. The main aim was to determine if the way they move could be related to their level of mental fatigue.

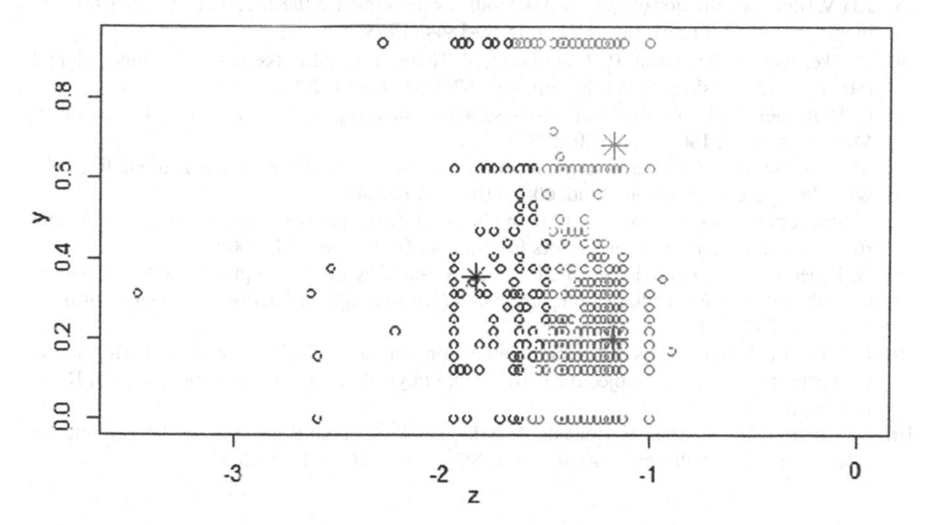

Fig. 6 Clusters for accelerometer 1

The results achieved from the implementation of the case study show that it is indeed possible to detect the onset of fatigue from the movement, caused most likely form the increased discomfort caused by mental fatigue. Moreover, the results also show that this approach, which is not intrusive concerning the user, is valid. The following limitations are acknowledged in the study. The number of participants was relatively small. Although it suffices as a proof of concept, we will be carrying out a larger study in the future. Moreover, asides from mental fatigue, questionnaires will also assess the level of comfort of the workers, which was not explicitly measured in this study.

While this approach may not be accurate enough on its own, it has the advantage of being based on relatively inexpensive sensors and of being non-invasive, especially when compared to existing approaches based on physiological sensors. Moreover, it can act as a valuable complement to existing fatigue monitoring systems.

Acknowledgments This work is part-funded by ERDF - European Regional Development Fund through the COMPETE Programme (operational programme for competitiveness) and by National Funds through the FCT (Portuguese Foundation for Science and Technology) within project FCOMP-01-0124-FEDER-028980 (PTDC/EEI-SII/1386/2012) and project PEst-OE/EEI/UI0752/2014.

References

1. P.V. Bulat, Evaluation of methods for the determination of factors inducing fatigue in man at work. Ergonomics **14**(1), 43–51 (1971)
2. M.H. Liao, C. Drury, Posture, discomfort and performance in a vdt task. Ergonomics **43**(3), 345–359 (2000)
3. J.H.V. Dieen, Evaluation of work-rest schedules with respect to the effects of postural workload in standing work. Ergonomics **41**(12), 1832–1844 (1998)
4. T. Åkerstedt, A. Knutsson, P. Westerholm, T. Theorell, L. Alfredsson, G. Kecklund, Mental fatigue, work and sleep. J. Psychosom. Res. **57**(5), 427–433 (2004)
5. E. Hollnagel, D.D. Woods, Cognitive systems engineering: new wine in new bottles. Int. J. Man Mach. Stud. **18**(6), 583–600 (1983)
6. M. Paradowski, A. Fletcher, Using task analysis to improve usability of fatigue modelling software. Int. J. Human Comput. Stud. **60**(1), 101–115 (2004)
7. R. Parasuraman, G.F. Wilson, Putting the brain to work: neuroergonomics past, present, and future. Hum. Factors: J. Hum. Factors Ergon. Soc. **50**(3), 468–474 (2008)
8. A. Pimenta, D. Carneiro, P. Novais, J. Neves, Monitoring mental fatigue through the analysis of keyboard and mouse interaction patterns, in *Hybrid Artificial Intelligent Systems* (Springer, 2013), pp. 222–231
9. L.P. Perelli, Fatigue Stressors in Simulated Long-duration Flight. Effects on Performance, Information Processing, Subjective Fatigue, and Physiological Cost. Technical Report, DTIC Document (1980)
10. M. Charrad, N. Ghazzali, V. Boiteau, A. Niknafs, NbClust: an R package for determining the relevant number of clusters in a data set. J. Stat. Softw. **61**(6), 1–36 (2014)

Development of a High Mobility Assistant Personal Robot for Home Operation

Eduard Clotet, Dani Martínez, Javier Moreno, Marcel Tresanchez and Jordi Palacín

Abstract This paper presents the development of an Assistant Personal Robotic (APR) designed with the objective of creating a high reliable robot that can be used in several home applications such as: home safety, elder people supervision and remote assistance, remote presence, etc. In this proposal the APR is remotely controlled by a smartphone or portable tablet with Wi-Fi connectivity. The APR design has taken into consideration safety factors, mobility and physical restrictions of an average home; including opened doors, tight turns, and narrow corridors. The APR design includes several onboard sensors in order to protect the robot and avoid collisions with fixed or moving surrounding objects.

Keywords Assistant personal robotic • Elderly care • Home assistant • Remote control system

1 Introduction

A home application such as supervising elderly people is an important task that does not use to require human presence. Multiple methods had been designed in order to keep contact with elderly people, such as performing daily calls or medical alert systems; those systems are very useful, but its efficiency relies on the consciousness of the user or the regularity of the calls. The effective supervision of elderly people at home can be performed by implementing sensor systems and methods designed to detect abnormal situations where elderly people are not able to call for assistance [1]. For example, in [2] a mobile platform equipped with infrared cameras is used in order to obtain the silhouette of the user and determinate its position: sitting in a chair, standing, or laying on the ground; even with the good

E. Clotet · D. Martínez · J. Moreno · M. Tresanchez · J. Palacín (✉)
Department of Computer Science and Industrial Engineering, University of Lleida, 25001 Lleida, Spain
e-mail: palacin@diei.udl.cat

A. Mohamed et al. (eds.), *Ambient Intelligence - Software and Applications*, Advances in Intelligent Systems and Computing 376,
DOI 10.1007/978-3-319-19695-4_7

results obtained, authors mentioned that the reliability of the system is compromised by the position of the camera. In [3] a novel system design for a remote-operated home robot is presented. The personal robotic assistant can also be used as a remote work station for people with restricted mobility, providing a more immersive and dynamic way to physically interact with other persons and its environment. In a similar direction, this paper proposes the development of a high mobility Assistant Personal Robot (APR) for home operation.

2 Materials and Methods

The main materials and methods used in the design of the APR can be classified as onboard electronics, high mobility system and remote control system.

2.1 High Mobility Capabilities

One of the most important features for a mobile robot is its ability to move around its surroundings as efficiently as possible without causing any damage to the environment or the robot itself. For this purpose, we propose the use of three omnidirectional wheels [4] in the design of the APR in order to allow displacements in any direction just by controlling the power applied to each one of the three DC motors that drives the omnidirectional wheels. Each wheel is rubber-covered in order to improve its traction in almost any surface.

The APR includes two symbolic soft arms also powered by two DC motors. These arms have only one degree of freedom and are designed mainly as an accessory element for fast non-verbal interaction or gesture imitation.

2.2 Central Processing Unit

The central processing unit (CPU) performs two main functions: manage all the communication with the remote control device and generate motion orders, and operate as a friendly interface in order to control the mobile robot and establish remote communication (audio and video) between the robot and a local user. A tablet device is used as the main CPU of the APR; the used device is the Asus Memo Pad 10″ tablet, which provides the basic elements required for remote supervision: two cameras, an embedded speaker, and a microphone. Other tablets with Android O.S. and similar specifications can also be used in the APR since the application is coded in Java, which is a high-level language. The tablet is then used to send, receive, and show data exchanged with the remote control device mainly through Wi-Fi technology.

The central processing unit is placed as the "head" of the APR which includes onboard DC motors in order to change the robot's head pan and tilt, similarly as a human head. This functionality simplifies the remote supervision of elderly people as the robot does not need to move when the remote supervision requires a different angle of view of the surroundings.

2.3 Onboard Electronics

Figure 1 shows the basic schema of the onboard electronics. The motor control unit (MCU) is the main electronic board of the APR, which receives the motion orders of the CPU and controls the different motors of the mobile robot. The MCU controls

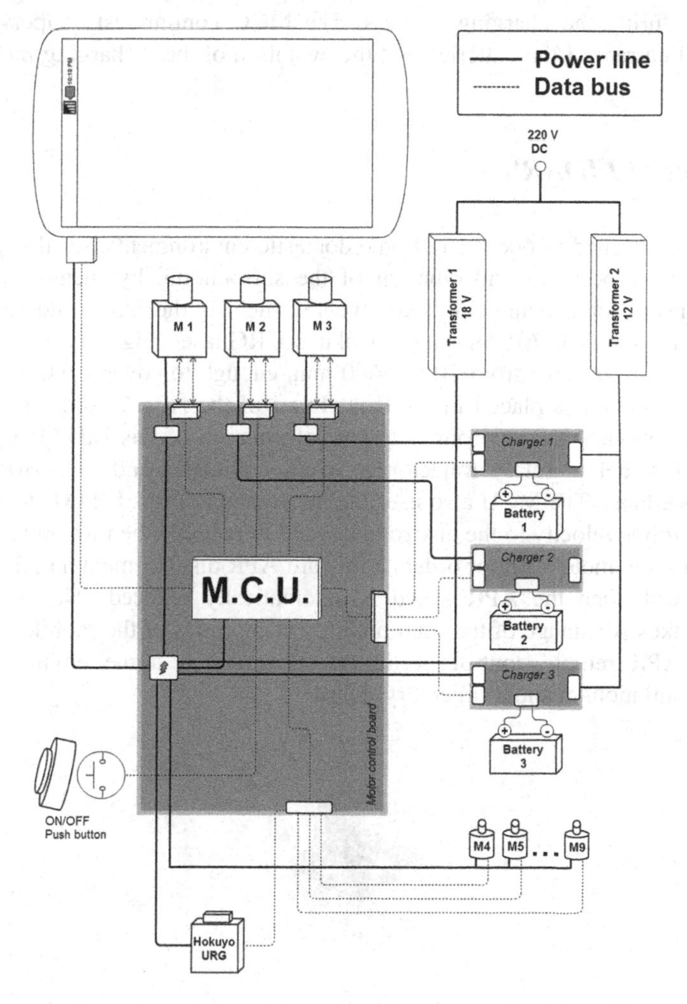

Fig. 1 Schematic draw of the onboard electronics

directly the different motors, batteries and safety sensors of the mobile robot. The MCU operates as a real-time device and has the responsibility of avoiding structural damage and collisions. The MCU supervises the status of the CPU in order to safely operate the APR in case of unexpected problems. The MCU is in charge of processing all the data obtained with the onboard LIDARS in order to ensure safe displacement paths in an unstructured home environment and avoid any collision. The mechanical design of the APR includes enough space to carry three batteries which are controlled by charging boards. These three batteries are individually connected to one of the main motion DC motors of the mobile robot. The APR is designed to work even during the charging process while it is plugged to an AC power socket. To this end, the APR includes three AC/DC power converters to charge individually each one of the batteries of the mobile robot and an additional AC/DC power converter that is capable of powering the MCU and ensuring the APR operability during the charging process. The MCU continuously supervises the voltage and current of the batteries and the evolution of the recharging procedure.

2.4 Onboard LIDARS

The APR is designed to operate in home domestic environments, for this purpose, the robot can process the information of the surrounding by means of several LIDAR sensors manufactured by Hokuyo [5] such as the URG and the UTM. Currently, the main LIDAR sensor is a Hokuyo URG laser (Fig. 2) which provides 240° of vision in a range from 20 to 5600 mm, enough to cover a whole room or corridor. The device is placed in the frontal part of the APR in order to obtain a planar description of the environment. The information of this LIDAR is used to detect objects, define safety trajectories, avoid collisions, and to perform basic SLAM procedures. The MCU also uses the information of this LIDAR to adequate the mobile robot velocity to the environment and to redefine the motion orders. For example, if the remote operator orders a forward APR displacement and the APR is near to a wall then the APR speed is automatically reduced. This supervised operation takes advantage of the high mobility capabilities of the mobile robot and simplifies APR remote control in tight spaces such a home environment by avoiding continuous manual trajectory adjust.

Fig. 2 Hokuyo URG LIDAR

2.5 APR-Tablet Application

The APR application running in the tablet of the mobile robot performs the following operations:

- The tablet of the APR is started when pressing the ON/OFF button of the mobile robot. In fact the ON/OFF button switches on the MCU and then the MCU powers the USB connection with the tablet. The APR-tablet has been configured to automatically switch on when detecting something connected to the USB port.
- Just after power-on, the APR-tablet detects a new USB device plugged and automatically starts a registered application program (APP) which is the conventional APR that controls all functions and capabilities of the mobile robot.
- The APR-APP automatically connects to a global server that manages all the client and robot connections, and announces itself as a robot ready to be controlled; after that, the application will wait for a client to connect (Fig. 3).
- Once a client is connected, the application starts a bidirectional videoconference and waits for incoming movement orders (Fig. 4).
- When a packet with a movement order is received, the application will convert it into motor orders and send it to the MCU through the USB port.
- Additionally, the MCU sends status information to the tablet that is submitted to the remote application in order to provide information of the mobile robot: battery status, emergency stop, etc.

Fig. 3 APR-tablet screen capture showing some images during a videoconference

2.6 Remote User Application

The remote application has to provide the following functions:

- Videoconference.
- APR remote displacement control.

The remote user application has been designed to be very intuitive. The remote user application performs the following operations:

- When started, this APP automatically tries to connect to a global server that manages the connections with the available mobile robots.
- The APP shows a list of all APR's available or configured for this remote client (Fig. 5).
- At this point the user only needs to click over the APR to control.
- Then, the application interface shows a layout that includes videoconference images and tactile mobile robot displacement controls (Fig. 6).

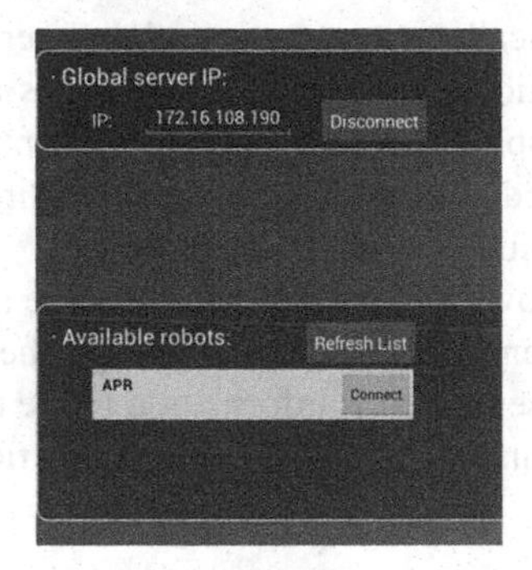

Fig. 4 Remote application showing robots available

Fig. 5 Remote application screen capture

3 APR Prototype Results

Figures 6, 7 and 8 show some APR prototype images. Figure 7 shows an image during a videoconference with a remote user. Figure 6 shows a detail of the omnidirectional wheels and the frontal LIDAR user for collision avoidance. The omnidirectional wheels allow tight maneuvers in reduced spaces such as: passing through doors, rotations, lateral displacements, etc. Finally, Fig. 8 shows the head of the mobile robot with the APR-tablet. The head has two degrees of freedom in order to control the angle of view without requiring a change in the position of the mobile robot, allowing to the robot to get visual information of any place in a one floor home.
The remote control application has been tested in multiple scenarios where the user connects to the robot from remote facilities, demonstrating the reliability of the remote control system. The APR has been remotely controlled from multiple locations with internet connection and the onboard LIDARS have avoided any collision with surrounding objects or elderly people around the mobile robot.

Fig. 6 Frontal view of the base with omnidirectional wheels

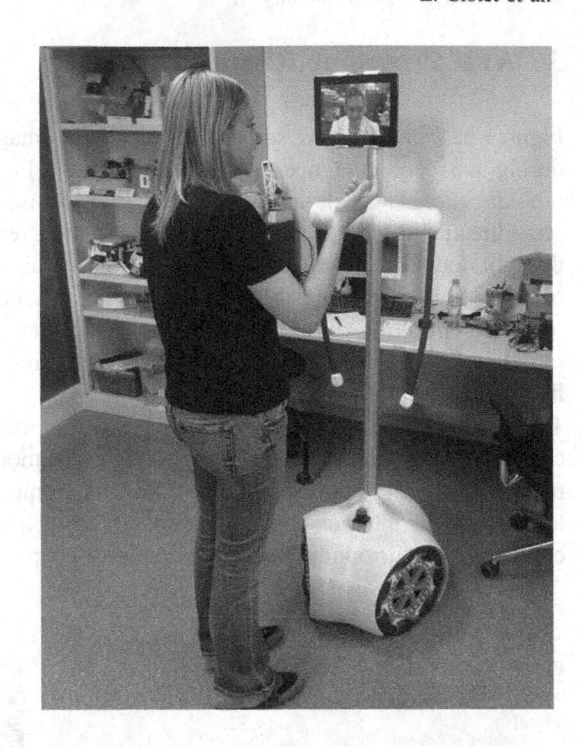

Fig. 7 APR overview

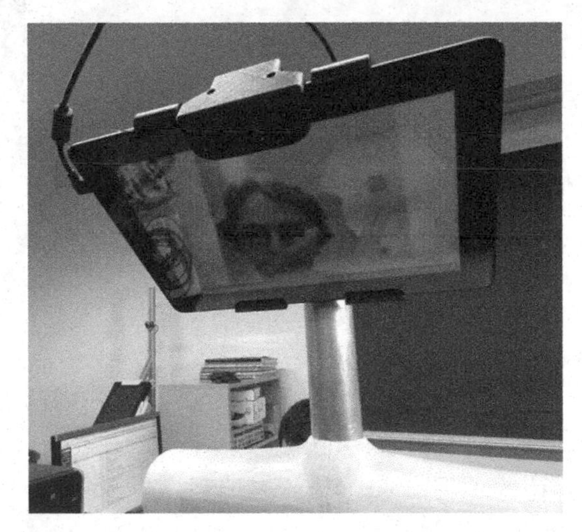

Fig. 8 Detail of the head of the APR

4 Conclusions

This paper proposes the implementation of an Assistant Personal Robot (APR) that can be used as a safety remote-presence system in order to monitor and assist elderly people in their home. The first obtained results showed that the proposed

APR prototype is able to overcome most of the physical barriers that can be found in domestic environments. Regardless its high mobility capabilities, the APR is not able to go through stairs or heavy slopes (greater than 37°), reducing its effectiveness in old buildings that are not adapted for people with reduced mobility, or buildings with multiple floors. The communication system has been reliable enough to work in LAN and WAN networks. Finally, the omnidirectional wheels have offered a great mobility and control over the robot by using the designed onscreen tactile remote control.

Acknowledgments This work is partially founded by Indra, the University of Lleida, and the RecerCaixa 2013 grant.

References

1. P. Pineau, M. Montemerlo, M. Pollack, N. Roy, S. Thrun, Towards robotic assistants in nursing homes. Robot. Auton. Syst. **42**(3–4), 271–281 (2003)
2. T. Liu, J. Liu, Mobile robot aided silhouette imaging and robust body pose recognition for elderly-fall detection. Int. J. Adv. Robot. Syst. **11**, (2013)
3. F. Michaud, P. Boissy, D. Labonté, H. Corriveau, A. Grant, M. Lauria, R. Cloutier, M.A. Roux, D. Iannuzzi, M.P. Royer, Telepresence robot for home care assistance. In Multidisciplinary Collaboration for Socially Assistive Robotics
4. J.-B. Song, K.-S. Byun, Design and control of a four-wheeled omnidirectional mobile robot with steerable omnidirectional wheels. J. Robot. Syst. **21**(4), 193–208 (2004)
5. D. Martínez, T. Pallejà, J. Moreno, M. Tresanchez, M. Teixidó, D. Font, T. Pardo, S. Marco, J. Palacín, A mobile robot agent for gas leak source detection. In 12th International Conference on Practical Applications of Agents and Multi-Agent Systems (PAAMS 2014), pp. 19–25, Salamanca, Spain (2014)

Using ICT for Tacit Knowledge Preservation in Old Age

Isabel Marcelino, José Góis, Rosalía Laza and António Pereira

Abstract As the world population is aging, numerous challenges were raised. How to maintain a sustainable aging? How to increase the active role of older adults in society? How to promote healthy aging along with the improvement of social and technological inclusion and enhance emotional well-being? How to preserve the vast tacit knowledge existent in seniors? Pervasive computing can giving an enormous contribution to overcome this issues. In the present paper we introduce eService platform as a novel service ecosystem mainly developed for senior population, including life experiences and knowledge record service. We have researched and selected the most relevant accessibility guidelines concerning senior population and made a low fidelity prototype, followed by both fidelity prototypes, one with and one without guideline application. Finally we conducted usability tests and semi structured interviews with 6 individuals to validate our work. The experimental results demonstrated that the proposed guideline checklist was validated, well accepted and easy to use by seniors. They also validate the extreme importance of knowledge preservation.

Keywords Gerontechnology · Knowledge preservation · Older adults

I. Marcelino (✉) · R. Laza
Higher Technical School of Computer Engineering, University of Vigo, Polytechnic Building, Campus Universitario as Lagoas s/n, 32004 Ourense, Spain
e-mail: isabel.marcelino@ipleiria.pt

R. Laza
e-mail: rlaza@vigo.es

I. Marcelino · A. Pereira
INOV INESC Innovation, Institute of New Technologies of Leiria, 2411-901 Leiria, Portugal
e-mail: apereira@ipleiria.pt

I. Marcelino · J. Góis · A. Pereira
School of Technology and Management, Computer Science and Communications Research Centre, Polytechnic Institute of Leiria, 2411-901 Leiria, Portugal
e-mail: joseldgois@hotmail.com

A. Mohamed et al. (eds.), *Ambient Intelligence - Software and Applications*,
Advances in Intelligent Systems and Computing 376,
DOI 10.1007/978-3-319-19695-4_8

1 Introduction

The world population is aging leading to several motivations [1]. These concerns include creating solutions to promote an active aging were senior can maintain their independence, dignity and contribution to society by playing an active role. Solutions to develop secure feelings in both elderly's and family members, since family member are in our days unable to give proper care to their older family. Solutions to grant that senior knowledge apprehended during a life time is not lost.

All of these challenges become even more complex due to the fact of having a heterogeneous group of population when concerning to elderly, since it might involve literate and unlettered individuals, with different interests and financial possibilities [2–4]. Nevertheless, some issues are common in aging. For instance vision decline, hearing loss, psychomotor coordination diminishment and cognitive deterioration. Older adults are also more likely to struggle with isolation, loneliness, and sadness feelings by not having an active role in society.

Having all of these concerns in mind, we can use available technologies to develop a solution that can minimize all the stated issues. The solution that our research team has developed is called eServices - Elderly Support Service Platform. eServices provides a novel service ecosystem mainly developed for senior population. It stands out from other approaches by providing a unique central access point to end users and by granting that interaction between end users and services is intuitive, accessible and easy to use. Our concern was also to compile several services that aimed both physical and psychological aspects. Yet, the service catalogue presented in eServices may be populated with external services granting scalability to our platform due to easy add-in services [5–7].

So far we have developed the platform, implement some services and defined a methodology to obtain information from the end users and available services integration combined with the basic life signs that allowed us to detect abnormal situations [8]. One of these services is focused on knowledge preservation. Knowledge preservation is an increasing concern as the global population is aging. Not only senior workers have built an organizational know-how and companies feel an impact when they retire, but also the knowledge of life experiences is being lost over generations, as long as family history information for clinical decision support. Additionally, there are benefits of storing memory that can trigger seniors to remember their past, which is very important in cognitive deterioration diseases, such as Alzheimer. Therefore, it is imperative to take measures that will allow wisdom safeguarding. The service life experiences and knowledge record existing in eServices is a measure to achieve it.

The purpose of the present paper is to give a contribution presenting and validating life experiences and knowledge record service.

The remainder of the paper proceeds as follows: related work is presented in Sect. 2, followed by the summarization of eServices platform in Sect. 3. Section 4 approaches the core of this paper concerning knowledge preservation. Section 5 contains the evaluation and results. Conclusions are presented in Sect. 6.

2 Related Work

As previously mentioned, knowledge preservation is a major concern in our days. Some solutions have been created to overcome this issue.

There are solutions that are more focused on record life events. Great life stories is one of these examples were a web page is available to share life chapters such as "Romance and Marriage", "Words of Wisdom", "Leisure and Travel" or "School Days" [9]. It is an interesting initiative, but it is not developed to meet a specific population, nor allows the inclusion of videos or images or even the ability to create more life chapters.

Another example is Our time lines [10]. In this web page it is possible to build a timeline where events has a type, a start and end date. But once again the solution is not developed specifically for a target population and is limited to textual information.

Microsoft Greenwich Project allows the creation of personal digital timelines and embebbed it on blogs. This tool is explored in [11] to stimulate older people memory.

The timeline format is also applied in one of the most popular social networks: Facebook.

Another project is [12], where members of this association help other people create personal histories, including memoirs, video tributes, autobiographies, biographies, family histories or cookbooks. It is interesting to observe some of the comments on this association's blog and see, not only senior people's will to leave their testimony, but also family members, especially sons, overjoyed to have their parents voices and laughs on video.

In [13], DVD-based multimedia biographies are built and observation is made over psychosocial effects that viewing the biographies had on the participants and their family members.

Digital life story scrapbook are also explored in people with dementia due to potential psychosocial benefits; although the existence of these benefits may be controversial as shown in the study presented in [14].

All of the presented approaches allow users to persist some of their life experiences and events and share it with others. Nevertheless, these solutions are not focused on a specific target population like seniors, who have particularities due to age-related issues. Moreover, older adults are a heterogeneous group where many of them are digitally illiterate or even uneducated. Therefore, a solution developed for this population segment must be aware of all these special needs to include as many users as possible. The developed solution must be simples, intuitive and follow accessibility guidelines that remove barriers between elderlies and Information Communication Technologies (ICT) usage. For this purpose we will briefly present eServices, the platform where life experiences and knowledge record service is nested.

3 Service Platform for Pervasive Elderly Care

eServices is a novel platform to aggregate several services directed to older adults. Its major and differentiating features are:

- Regarding to seniors covering not only the physical, but also social and emotional part. It provides basic life signs monitoring as long as a virtual room were other services like medical, leisure are available;
- Extreme simplicity with special attention to the interface designed according to accessibility guidelines to suppress usage barriers from illiteracy and technological illiteracy;
- Non-invasive and integrated in daily basic activities;
- Reliable and preventive by creating behavior analysis that combine both motorization of vital signs and interaction between the user and platform, therefore reducing false alerts in detecting abnormal situations;
- Grant immediate response in imminent danger situations by providing several alert levels, personalized for each individual;
- Secure and considering ethical issues by avoiding quasi-identifiers and protect user identities from service providers and ciphering sensitive information;
- Providing a single access point for users to obtain their services, scaling new services in order to promote a platform with integration and interoperability;
- Designed with pervasive computing concern: any service, anytime, anywhere, any device, any access, any people…, AnyN;
- Offering a modular solution that can be achieved, in a production phase, by elderlies according to their financial possibilities. For instance, they can purchase the services that they are more interested. Concerning to physical monitoring, sensors where also built in modular and scalable way. Basic sensors grant the ability to provide basic sensing and modules may be copulated to acquire specific parameters for certain user.

eServices as a modular architecture comprising of 3 major components: end users clients, services providers and the middleware platform between end users and service providers.

End user client comprises biosensors to monitor basic life signs, environment sensors and service access device to interact with the provided services.

The middleware platform contains service catalogue, system administration and event management.

Service providers will have to register their service in one of the existing categories in service catalogue or ask to create a new one. Service providers will also need to use middleware platform communication protocol.

Some categories were already identified: medical, maintenance, call center, leisure and cultural. Life experiences and knowledge record is one of the services available in leisure category and will be approached in the next section.

4 Knowledge Preservation

As previously mentioned, knowledge preservation is an emerging concern as the experience gained over a lifetime often dies along with people, not passing to forthcoming generations. Life experiences and knowledge record is a service available in eServices platform. Its main goal is to provide the ability for older adults, with or without technological skills, to record, view and share their life stories with other users of eServices. In this information we comprehend handcrafts, recipes, folk medicine, proverbs, traditional agriculture methods, stories to grandchildren, etc.

As a service, the only requirements that elderly must have is a device with internet connection and browser. The absence of complex installation is crucial regarding elderly population. Another vital aspect of this service was to develop a web site that meet accessibility standards to take into account age-related physical changes and cognitive impairments. Some of the World Wide Web Consortium (W3C) recommendations and guidelines were applied [15–17]. From the recommendations provided by W3C, we have collected a subset that we considered essential when developing interfaces for elderlies and applied them to our prototype, namely:

- Maintain contrast due to vision issues usually experienced in old age;
- Avoid patterned backgrounds preferring the use of solid colors;
- Use text fonts easy to read, namely sans-serif text types and non-italic, and always allied with audio and representative icons to grant the perception by all users;
- Keep to number of steps to perform an action as reduced as possible to avoid excessive memorization;
- Give an alternative to interact to the computer by other means than traditional inputs such as keyboards and mouse, using voice commands and touchscreen interfaces;
- Remove scrolls that will increase navigation complexity;
- Maintain a consistent layout, granting uniformity;
- Be aware that older users take more time to consolidate information. It is important to give them the proper time to accomplish certain task;
- Use concise language, short and clear messages without ambiguity;
- Use large icons that are easy to click.

A low fidelity prototype was firstly developed followed by 2 high level prototype. One of the final prototypes was developed with the stated guidelines and the other without these guidelines. The implementation as made using HTML5 and YouTube Data API as long as YouTube Upload Widget [18].

The purpose was to evaluate the impact of using specific guidelines for senior population as long as validate the importance of knowledge preservation service.

5 Evaluation and Results

Regarding evaluation and methodology, as stated above, a low fidelity prototype was firstly developed and discussed with several seniors in order to fully align their expectations towards the final solution. After validation, 2 high level prototype were developed (one including the accessibility guidelines that our research team as found most valuable for senior people, other without these guidelines). We have conducted usability tests and semi structured interviews with 6 individuals. The tests and interviews were conducted in the individual's environment to reduce any interference from being in a strange place.

Before the usability tests, our research team members have explained how the system works and made a brief demonstration. Afterwards, we have made some control questions that can be observed in Tables 1 and 2. These questions were made over conversation to grant that elderlies were not anxious when realizing the tests. Next we have conducted usability tests by asking the users to accomplish the tasks specified in Table 3 in both prototypes. Finally we interviewed them to gather their general opinion on the experience and the significance of having life experiences and knowledge record service.

Table 1 Control questions

Q1	Personal data: age and sex
Q2	Former life occupation and current interests (handcraft, fishing, other)
Q3	Iliteracia
Q4	ICT know-how: have ever used mobile phone, computer, and internet?
Q5	Health issues (vision, hearing, motor, cognitive)

Table 2 Control questions results

<table>
<tr><th>Q1</th><th>Q2</th><th>Q3</th><th>Q4</th><th>Q5</th></tr>
<tr><td>42 Male</td><td>Active farm worker | travels and agriculture</td><td>No</td><td>Uses a mobile phone; don't have any computer skills</td><td>Vision issues, manifesting myopia</td></tr>
<tr><td>65 Male</td><td>Retired construction worker | someone to talk to</td><td>No</td><td rowspan="5">Technologically illiterate; none of them have any computer skills or interaction or ever had used a mobile phone</td><td>Vision issues hearing problems</td></tr>
<tr><td>76 Female</td><td>Retired farmer | someone to talk to, continue to plant and crochet handicraft</td><td>Yes</td><td>Vision issues hearing problems</td></tr>
<tr><td>80 Female</td><td>Retired farmer | someone to talk to, continue to plant</td><td>No</td><td>Vision issues</td></tr>
<tr><td>83 Female</td><td>Retired farmer | someone to talk to, continue to plant and crochet handicraft</td><td>Yes</td><td>Vision issues hearing problems tremors</td></tr>
<tr><td>84 Female</td><td>Retired farmer | someone to talk to</td><td>Yes</td><td>Vision issues Alzheimer in an earlier stage</td></tr>
</table>

Table 3 Tasks

T1	Record video
T2	Watch recorded video
T3	Share video
T4	Watch my videos
T5	Watch videos from a specific channel

For each task stated in Table 3, we have verified, for both prototypes, which were successfully performed. We have made an evaluation over a 5 point rating scale. For this classification we have considered the time to learn, user errors and subjective satisfaction. In the results presented in Figs. 1 and 2 it can be concluded that the prototype with the accessibility guidelines have a better rate.

In interviews made the elderlies demonstrated to be overjoyed with the possibility of recording their memoires and know-how, sharing it with their friends and family, namely grandchildren, especially those who had family abroad.

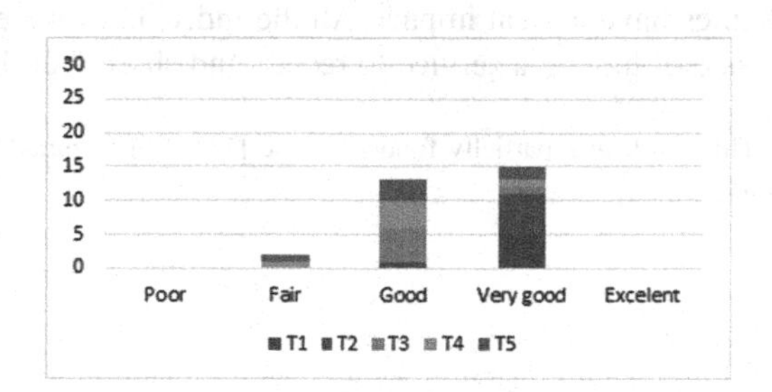

Fig. 1 Results in evaluating the prototype with accessibility guidelines applied

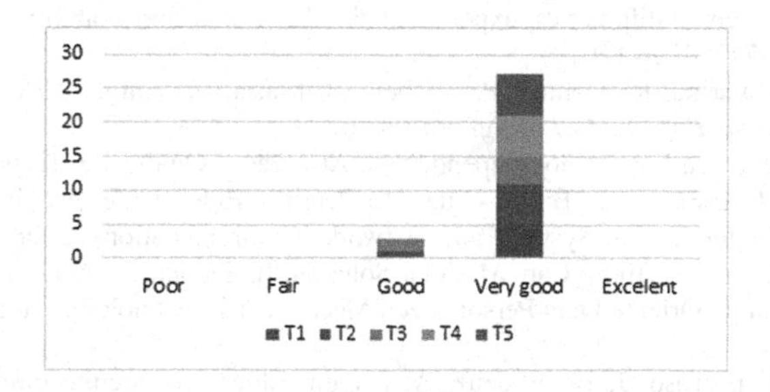

Fig. 2 Results in evaluating the prototype without accessibility guidelines applied

6 Conclusions

It is a known fact that world life expectancy is increasing and that elderlies can improve their lives benefiting of ICT services. One of the major concerns in having an aging population is knowledge preservation. This concern is shared by several stakeholders. For elderly to revisit the past, give a contribution to prosperity and help trigger long-term memories of individuals with dementia; For family members allowing them to remember dear family members that have passed away, to maintain their life memories and knowledge over generations, to have a clinical history that can provide useful medical information to present generations; For companies to maintain their know-how created by tacit knowledge of their retired employees.

This paper presents a solution to improve elderly's quality of life and well-being by promoting knowledge preservation. The solution was to build a scalable service platform and developing life experiences and knowledge record service. The key features during the conception and development was to keep simplicity, involve the target population in the solution design, identify and apply accessibility guidelines essential to senior population. A low fidelity prototype was made. Afterwards, by 2 high fidelity prototype were made, one with the selected accessibility guidelines and one without. The solution was validated over usability tests and semi structured interviews to 6 individuals. The obtained results univocally showed that the chosen accessibility guidelines have a great impact. All the individuals were unanimous in stating the importance of having a service to record and share their life experience.

Acknowledgments This work was partially funded by the [14VI05] Contract-Programme from the University of Vigo.

References

1. U. Nations, Population Ageing and Development 2012 (2012)
2. Y. Barnard, M.D. Bradley, F. Hodgson, A.D. Lloyd, Learning to use new technologies by older adults: perceived difficulties, experimentation behaviour and usability. Comput. Human Behav. **29**(4), 1715–1724 (2013)
3. V.T. Oliveira Catarina, R. Manuel, P. Anabela, B. Maria, M. António, *Estudo do Perfil do Envelhecimento da População Portuguesa* (2010)
4. O Envelhecimento da População: Dependência, Ativação e Qualidade (Lisbon, 2012)
5. I. Marcelino, J. Barroso, J. Bulas Cruz, A. Pereira, Elder Care Architecture. In Third International Conference on Systems and Networks Communications, 2008, pp. 349–354
6. I. Marcelino, A. Pereira, Elder Care Modular Solution. In Second International Conference on Advances in HumanOriented and Personalized Mechanisms Technologies and Services, 2009, pp. 1–6
7. I. Marcelino, J. Barroso, J. Bulas Cruz, A. Pereira, Elder care architecture, a physical and social approach. Int. J. Adv. Life Sci. **2**(61.2), 53–62 (2010)
8. I. Marcelino, D. Lopes, M. Reis, F. Silva, R. Laza, A. Pereira, Using the eServices platform for detecting behavior patterns deviation in the elderly assisted living : a case study. Biomed. Res. Int. (2014)
9. Great Life Stories (2007), http://www.greatlifestories.com/. Accessed 04 May 2013
10. Our Time Lines, http://www.ourtimelines.com/index.shtml
11. E. Thiry, S. Lindley, R. Banks, T. Regan, Authoring personal histories : exploring the timeline as a framework for meaning making (2013)
12. Association of Personal Historians (2011), http://www.personalhistorians.org/. Accessed: 04 May 2013

13. M. Crete-Nishihataa, R. Baeckera, M. Massimia, D. Ptaka, R. Campigottoa, L. Kaufmana, A. Brickmanb, G. Turnerc, J. Steinermand, S. Blackc, Reconstructing the past: personal memory technologies are not just personal and not just for memory. Human–Comput. Interact. (2012)
14. P. Subramaniam, B. Woods, The impact of individual reminiscence therapy for people with dementia: systematic review. Expert Rev. Neurother. **12**(5), 545–555 (2012)
15. H. Petrie, N. Bevan, The evaluation of Accessibility, Usability and User Experience. In The Universal Access Handbook, G. Constantine Stephanidis, Foundation for Research & Technology—Hellas, Institute of Computer Science, Crete, Greece and University of Crete, Department of Computer Science, Crete, Ed. (2009)
16. A. Arch, S. Abou-Zahra, H. Shawn, Older users online: WAI guidelines address older users web experience. User Exp. Mag. **8**(1), (2009)
17. W3C Standards—Accessibility (2013), http://www.w3.org/standards/webdesign/accessibility. Accessed: 18 May 2014
18. Youtube for Developers, https://developers.google.com/youtube/

Step Count and Classification Using Sensor Information Fusion

Ricardo Anacleto, Lino Figueiredo, Ana Almeida, Paulo Novais and António Meireles

Abstract In order to suppress the GNSS (Global Navigation Satellite System) limitation to track persons in indoor or in dense environments, a pedestrian inertial navigation system can be used. However, this type of systems have huge location estimation errors due to the Pedestrian Dead Reckoning (PDR) characteristics and the use of low-cost inertial sensors. To suppress some of these errors we propose a system that uses several sensors spread in person's body combined with information fusion techniques. Information fusion techniques provide lighter algorithms implementations, to count and classify the type of step, to run in mobile devices. Thus, improving pedestrian inertial navigation systems accuracy.

Keywords Pedestrian inertial navigation system · Step count · Information fusion · Dynamic time warping

This work is part-funded by ERDF - European Regional Development Fund through the COMPETE Programme (operational programme for competitiveness) and by National Funds through the FCT - Fundação para a Ciência e a Tecnologia (Portuguese Foundation for Science and Technology) within project FCOMP-01-0124-FEDER-028980 (PTDC/EEI-SII/1386/2012). Ricardo also acknowledge FCT for the support of his work through the PhD grant SFRH/DB/70248/2010.

R. Anacleto (✉) · L. Figueiredo · A. Almeida · A. Meireles
GECAD—Knowledge Engineering and Decision Support Research Center,
School of Engineering—Polytechnic of Porto, Porto, Portugal
e-mail: rmsao@isep.ipp.pt

L. Figueiredo
e-mail: lbf@isep.ipp.pt

A. Almeida
e-mail: amn@isep.ipp.pt

A. Meireles
e-mail: ajmme@isep.ipp.pt

P. Novais
CCTC—Computer Science and Technology Center, University of Minho, Braga, Portugal
e-mail: pjon@di.uminho.pt

A. Mohamed et al. (eds.), *Ambient Intelligence - Software and Applications*,
Advances in Intelligent Systems and Computing 376,
DOI 10.1007/978-3-319-19695-4_9

1 Introduction

A system that is capable to locate an individual can be explored to improve life quality since emergency teams (fire-fighters, military forces [5] and medics) can respond more precisely if the team members location is known, tourists can have better recommendations [8], the elderly can be better monitored [9] and parents can be more relaxed with their children in shopping malls [3].

Usually these applications retrieve pedestrian's location by using a GNSS (Global Navigation Satellite System). Unfortunately, GNSS signals aren't available inside buildings or in dense environments. Consequently location-aware applications sometimes cannot know the user location.

There are already some proposed systems that retrieve location in indoor environments. However, most of these solutions require a structured environment [6]. Therefore, these systems could be a possible solution for indoor environments, but in a dense forest or urban canyons they are difficult to implement.

To suppress structured environment limitations, a Pedestrian Inertial Navigation Systems (PINS) can be used. Typically, a PINS is based on an algorithm that involves three phases: step detection, step length estimation and heading estimation. A PINS uses accelerometers, gyroscopes, among other sensors, to continuously calculate via dead reckoning the position and orientation of a pedestrian. These sensors are based on MEMS (Microelectromechanical systems), which are tiny and lightweight sensors making them ideal to be integrated into the person's body or clothes. Unfortunately, large deviations of inertial sensors can affect these systems performance, so the PINS big challenge is to correct the sensors deviations.

In the research team previous works, the step detection was improved by using an algorithm that combines an accelerometer and force sensors placed on the pedestrians foot [2]. Then this led to better results [1] on the estimation of the pedestrian displacement. However, it still exists an error of 0.4 % in step detection and an error of 7.3 % in distance estimation.

We have found that a PINS solution only based on one IMU (Inertial Measurement Unit), composed by an accelerometer and a gyroscope, is not accurate enough. Thus we believe that using several IMU in the person's body, combined with a information fusion strategy, will improve the accuracy of PINS.

This goal is addressed throughout the document, where the system architecture is presented in Sect. 2 and the developed algorithms, that detect pedestrian steps and classify them, are presented in Sect. 3. In Sect. 4 the experimental results are given and, finally, in Sect. 5 are discussed the conclusions and the future work.

2 System Architecture

The proposed system is composed by two low-cost IMU, developed by the authors [2], and an "Integration Software" (described in Sect. 3) that integrates the information from the IMUs to count and classify the pedestrian steps. The system architecture is demonstrated in Fig. 1.

When referring to a low-cost IMU it implies different things for researchers, since for some a thousand euros IMU is considered low-cost. However, for PINS a low-cost IMU should cost less than € 100. This price restriction, implies the use of MEMS sensors that are truly low-cost.

The first IMU (Waist IMU), presented in Fig. 2a, is placed on the abdominal area and is composed by a STMicroelectronics L3G4200D gyroscope [12], a Analog Devices ADXL345 accelerometer [4] and a Honeywell HMC5883L magnetometer [7]. The second IMU (Foot IMU) is placed on the foot and is presented in Fig. 2b. It is composed by an Analog Devices ADXL345 accelerometer [4], a STMicroelectronics L3G4200D gyroscope [12] and two Tekscan FlexiForce® A201 force sensors [13]. The accelerometer is used to detect and quantify the foot movement, and the gyroscope is valuable to transform the acceleration data from the sensor frame to the navigation frame.

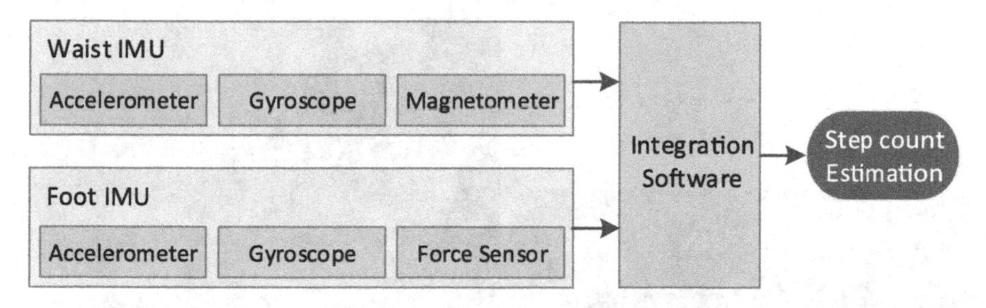

Fig. 1 System architecture

Force sensors were included since they can improve the process of detection of the moment when the user touches his feet on the ground, as well as, the correspondent contact force, which combined with the accelerometer improve the accuracy of the step length estimation. One force sensor was placed in the front part of the foot and the other in the heel, as shown in Fig. 2b.

The IMU collects the data with a rate of 100 Hz. However, the mean of the last five readings is made to reduce some of the errors, meaning that the data rate decreases to 20 Hz, which is sufficient to include the signal frequencies induced by the walking of a pedestrian.

Next section will present the "Integration Software" which classifies the step.

3 Step Count and Classification Algorithms

From previous experiences the step detection and classification using only an accelerometer or an accelerometer and a force sensor, still has some error.

In Fig. 3a is represented an acceleration signal obtained from the foot accelerometer, for a backward step. It can also be seen a simulated acceleration signal, which is the one expected when a backward step is given. As can be seen the accelerometer doesn't capture the accelerations in a perfect form, but it contains the information needed to be used to classify a step.

Although the pattern of the acceleration can be used to classify a step, sometimes the accelerometer produce a signal that doesn't follow any pattern, which turns to be useless to correctly classify a step. In Fig. 3b is represented an acceleration signal, that had occurred in a forward step, that doesn't follow any pattern. This acceleration signal can't be used to correctly classify a step.

Using several sources of data can be useful to surpass some of these random readings. The probability of two sources of data give erroneous acceleration patterns at the same step is very reduced. The fusion between all the sensors information can improve the number of correct classifications. However, the integration of more than one IMU can be very difficult to implement.

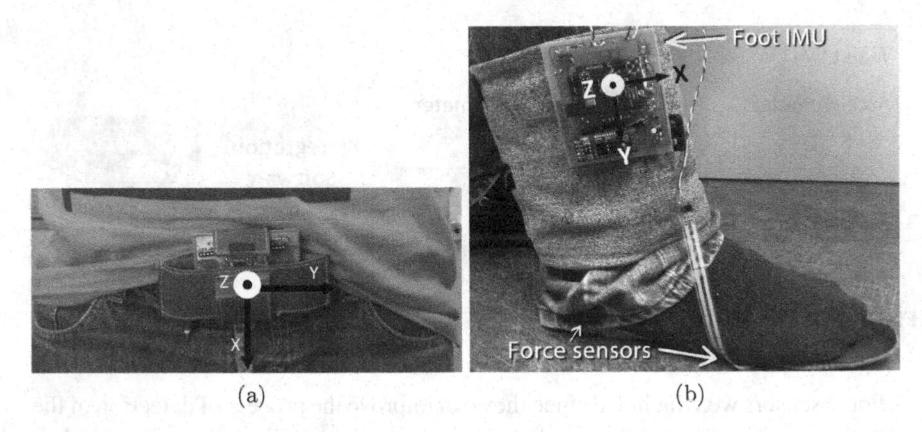

Fig. 2 Foot (a) and waist (b) IMU with the corresponding axis

The objective of this proposal is to detect and classify a step given by a pedestrian, combining several sources of information. After the detection of a step, with the algorithm explained in [1] and in [2], the proposed algorithm classifies the step as forward or backward.

Three algorithms were implemented to classify the direction of a step. The first one is based on some heuristics (Sect. 3.1). The second is based on a Dynamic Time Warping (DTW) [10] approach (Sect. 3.2). Both use only the data of one IMU. The third one (Sect. 3.3) uses a heuristic approach and a weight fusion technique to combine the data from the two IMU, to achieve a consensus about the characterization of the step.

Due to the existence of more than one IMU, an important issue is the data time synchronization. During data collection both IMU sensor data must be synchronized with an accuracy sufficient for this type of application. In our case, when the foot IMU data is received by the waist IMU, a timestamp is assigned to the data.

3.1 Heuristic Method

Typically a PINS detects a step by using the accelerometer data and by analyzing the forward and upward accelerations during the walking path. Typically, the detection is performed by using at least one of these three methods: peak detection, zero crossing detection and flat zone detection. Analyzing the literature can be seen that the peak detection is the most used method. However, peak and zero crossing detection algorithms can miss or over detected some steps because of accelerometer erroneous signal.

This sensor signal also shows distinguishable characteristics for walking characterization.

After smoothing the raw acceleration data and to classify the direction of the step, in both IMU, the Eq. 1 was used.

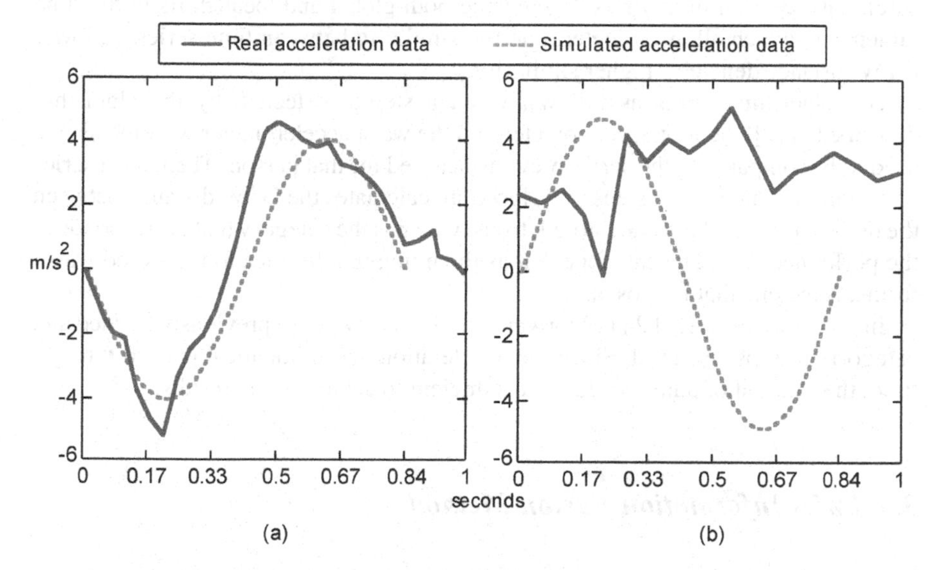

Fig. 3 (a) Acceleration data sensed, by the foot accelerometer, in a backward step; (b) erroneous acceleration data sensed, by the foot accelerometer, for a forward step

$$\sum_{i=1}^{\frac{n}{2}} acc_i > \sum_{i=\frac{n}{2}}^{n} acc_i \tag{1}$$

where n represents the number of acceleration values detected for the step, acc is the acceleration values sensed on the x axis, in the case of the foot IMU, and on the z axis, in the case of the waist IMU.

This formula sums the first half of the signal and compares it with the sum of the second half. If the first is positive or higher than the second it is a forward step, if not it is a backward step. From our tests the maximum number of acceleration samples for a step was 60, meaning that this algorithm is fast to process.

3.2 Dynamic Time Warping Method

DTW is a well-known technique, in time series analysis, which finds an optimal alignment between two given time series. It is mainly used to measure the similarity between two temporal sequences which may vary in time or speed. The main problem is that the DTW algorithm has a quadratic, $O(n^2)$, time and space complexity that limits its use to only small time series data sets. It gives intuitive distance measurements between time series by ignoring both global and local shifts in the time dimension, which allows to determine the similarity between time series. A lower DTW distance denotes a higher similarity.

Our algorithm works as follows, when a step is detected, by the algorithm described in [2], the foot accelerometer and the waist accelerometer waveforms, are used to be compared to the series previous learned for that person. Then, for a series of previously categorized steps, the algorithm calculates the DTW distance between the detected step and the ones stored, to see which is the category that corresponds to the performed step. The category that as the minimum distance to the stored waveforms, is the one that is chosen.

In our tests a dataset of 24 (12 forward and 12 backward) of previously learned and categorized steps was used. Since the acceleration has an identical pattern through time, this amount of data proved to be sufficient to achieve good results.

3.3 IMUs Information Fusion Method

The two IMUs placed in the person's body allows to combine the information given by them in order to minimize the complexity of the algorithms and maximize the accuracy and the robustness of the navigation solution.

Typically there are three types of fusion: data fusion, feature fusion and decision fusion. In this case the decision fusion was chosen. For each source of data, foot accelerometer and waist accelerometer, it is calculated the probability of the predicted result. This probability is calculated, according to Eq. 2, which is based on the fact that a positive acceleration must be followed by a negative acceleration of the same magnitude, and vice-versa.

$$stepprobability = 100 - (abs((max(acc) + min(acc))) \times 20) \tag{2}$$

If the acceleration signal doesn't follow this pattern then a low probability is given to it. For the acceleration example shown in Fig. 3a the probability that it is a characterizable step is 100 % and for the example shown in Fig. 3b the probability is only 20 %.

After the calculation of this probability, a weight is given to each one of the data sources. The foot IMU has a weight of 0.6, since it is the most reliable source of data, and the waist IMU has a weight of 0.4.

4 Experimental Results

The developed system and algorithms were evaluated by using a dataset of 200 steps performed by two pedestrians (100 steps for each pedestrian). The data was collected and then post processed using Matlab to obtain the results, meaning that the same dataset was used to test each algorithm.

The test scenario is a straight walk with two 90° turns, in the middle of the path, one to the left and the other to the right. A total of 25 steps (13 forward and 12 backward), each time, were performed in this scenario which gives a total traveled distance of 10 m and a displacement of 5 m. Four runs in this scenario, for each pedestrian, were performed.

The obtained results can be seen in Table 1. This table presents for each algorithm, the categorization accuracy (in percentage) for each IMU and the execution time (in milliseconds). The simulations were performed on a low performance computer, a Pentium 4 2.8 Ghz with 1 GB of RAM memory.

Table 1 Results for the three implemented algorithms

Method	Forward		Backward		Execution time
	Waist IMU	Foot IMU	Waist IMU	Foot IMU	
Heuristics	82.1 %	97.4 %	74.4 %	94.9 %	1 ms
DTW	84.6 %	100 %	79.5 %	97.4 %	100 ms
IMUs information fusion	100 %		98 %		2 ms

From the obtained results, it can be concluded that the waist IMU produces more errors than the foot IMU. This mainly happens because when the user is moving the foot is a more stable platform than the waist. A lot of unwanted accelerations are sensed by the waist, which leaves to a poor characterization of the step, but there are some features that can be retrieved to help other sources to properly characterize the step.

Regarding the step characterization the backward one is more difficult to classify than the forward one. Mainly because of the errors, presented on Sect. 2, that can occur in the accelerometer readings.

Comparing the DTW approach with the Heuristic one, it can be seen that the DTW has lower errors. However, it has an execution time 100 times longer than the Heuristic one. In order to maintain a lower execution time and an accuracy similar to the DTW approach, the information of both IMUs was fused. Through the sensors complementarity the step was categorized with similar accuracy but with an execution time 50 times smaller. This is an important help in order to improve the pedestrian displacement estimation. Using IMUs in different locations on pedestrian body, waist and foot, was very important to have these results.

5 Conclusion

Develop an accurate, inexpensive and small PINS to be used by persons, when they are on foot can be a huge challenge. Many approaches already have been proposed, but must of them rely on a structured environment that usually is unfeasible to implement and the other's don't provide the necessary accuracy.

In this work two IMUs were used, one on the foot and the other on the waist, where their data was explored to the maximum in order to provide an acceptable level of performance. Since the detection of stance phase using only accelerometers can introduce several errors on PINS, our proposal uses information fusion techniques to improve step detection and its classification. Through the use of these techniques an average accuracy of 99 % was achieved, which is very satisfactory.

In the future we want to use this step classification to improve distance estimation. Also, we want to use different estimation algorithms for each state, forward or backward, because it is more natural for a human to perform a forward step than a backward one. Meaning that the patterns for a forward step are more constant than for a backward step, since it isn't natural to us do that type of movement.

References

1. R. Anacleto, L. Figueiredo, A. Almeida, P. Novais, Localization system for pedestrians based on sensor and information fusion, in *17th International Conference on Information Fusion (FUSION)*, pp. 1–8, July 2014
2. R. Anacleto, L. Figueiredo, A. Almeida, P. Novais, Person localization using sensor information fusion, in *Ambient Intelligence—Software and Applications*, ed. by C. Ramos, P. Novais, C.E. Nihan, J.M. Corchado Rodrguez. Advances in Intelligent Systems and Computing, vol. 291. (Springer International Publishing, 2014), pp. 53–61
3. R. Anacleto, N. Luz, A. Almeida, L. Figueiredo, P. Novais, Shopping center tracking and recommendation systems, in *Soft Computing Models in Industrial and Environmental Applications*, in *6th International Conference SOCO 2011, in Advances in Intelligent and Soft Computing*, vol. 87. (Springer, Berlin Heidelberg, 2011), pp. 299–308

4. Analog Devices, Adxl345 digital accelerometer (2014)
5. J. Elwell, Inertial navigation for the urban warrior. Proc. SPIE **3709**, 196–204 (1999)
6. J. Hightower, G. Borriello, Location systems for ubiquitous computing. IEEE Comput. **34**(8), 57–66 (2001)
7. Honeywell, 3-axis digital compass ic hmc58831 (2014)
8. J. Lucas, N. Luz, M. Moreno, R. Anacleto, A. Almeida, C. Martins, A hybrid recommendation approach for a tourism system. Expert Syst. Appl. **40**(9), 3532–3550 (2013). July
9. J. Ramos, R. Anacleto, Â. Costa, P. Novais, L. Figueiredo, A. Almeida, Orientation system for people with cognitive disabilities, *Ambient Intelligence—Software and Applications*, *in Advances in Intelligent and Soft Computing*, vol. 153. (Springer, Berlin Heidelberg, 2012), pp. 43–50
10. D. Sankoff, J. Kruskal, *Time Warps, String Edits, and Macromolecules: The Theory and Practice of Sequence Comparison* (Center for the Study of Language and Inf, Stanford, 1999). December
11. J. Saunders, V. Inman, H. Eberhart, The major determinants in normal and pathological gait. J. Bone Joint Surg. **35**(3), 543–558 (1953)
12. STMicroelectronics, L3g4200d: three axis digital output gyroscope (2014)
13. Tekscan, Flexiforce sensors for force measurement (2014)
14. C. Vaughan, B. Davis, J.C. O'Connor, *Dynamics of Human Gait* (Human Kinetics Publishers Champaign, Illinois, 1992)

ECG Signal Prediction for Destructive Motion Artefacts

António Meireles, Lino Figueiredo, Luís Seabra Lopes
and Ricardo Anacleto

Abstract This paper addresses the ability of Burg algorithm to predict the ECG signal when it was completely destroyed by motion artefacts. The application focus of this study is portable devices used in telemedicine and healthcare, where the daily activity of patients produces several contact losses and movements of electrodes on the skin. The paper starts with a short analysis of noise sources that affects the ECG signal, followed by the algorithm implementation and the results. The obtained results show that Burg algorithm is a very promising technique to predict the ECG signal for at least three sequential heart beats.

Keywords Linear prediction · ECG · Burg algorithm · e-Health

This work is part-funded by ERDF - European Regional Development Fund through the COMPETE Programme (operational programme for competitiveness) and by National Funds through the FCT Fundação para a Ciência e a Tecnologia (Portuguese Foundation for Science and Technology) within project FCOMP-01-0124-FEDER-028980 (PTDC/EEI-SII/1386/2012). António also acknowledge FCT for the support of his work through the PhD grant (SFRH/BD/62494/2009).

A. Meireles (✉) · L. Figueiredo · R. Anacleto
GECAD—Knowledge Engineering and Decision Support Research Center,
School of Engineering—Polytechnic of Porto, Porto, Portugal
e-mail: ajmme@isep.ipp.pt

L.S. Lopes
IEETA—Institute of Electronics and Telematics Engineering of Aveiro, Aveiro, Portugal
e-mail: lsl@ua.pt

L. Figueiredo
e-mail: lbf@isep.ipp.pt

R. Anacleto
e-mail: rmsao@isep.ipp.pt

A. Mohamed et al. (eds.), *Ambient Intelligence - Software and Applications*,
Advances in Intelligent Systems and Computing 376,
DOI 10.1007/978-3-319-19695-4_10

1 Introduction

It is well known that an electrocardiogram (ECG) is a representation of the heart electrical activity in order to time. The ECG signal is of considerable importance since according to the recent statistics report published by World Health Organization, cardiovascular diseases remaining the main reason of mortality in the world [13]. ECG signal is composed by a P wave and a QRS complex corresponding to the heart depolarization, and T wave corresponding to heart repolarisation. A normal ECG shape can be seen in Fig. 1.

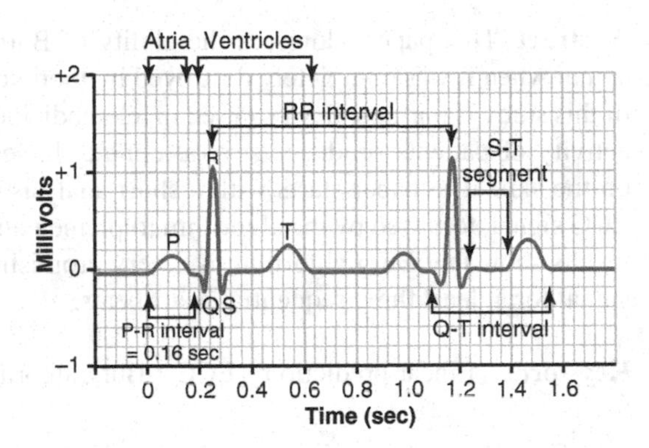

Fig. 1 Normal electrocardiogram [6]

Nowadays, the use of portable heart monitoring devices is very common. These devices are an excellent resource for ambulatory ECG and patient monitoring for long periods of time. Unfortunately, the constant movement of the human body produces several artefacts, caused by the movement of the electrodes in the skin. This noise is usually known as motion artefacts and it is a kind of noise that modifies or shifts the ECG signal or completely destroys it. The main goal of this work is to test and validate a technique to detect these artefacts, and based on the previous ECG signal, create a signal with a linear prediction method based on Burg Algorithm. The presented experiments also have relevance due to the fact of this algorithm does not have been well explored for this subject.

2 Biomedical Noises Sources

ECG signals always have background noise associated, such as power line interferences, muscular contractions (electromyography, EMG), or instrumentation noise generated by electronic devices. But, in a normal functioning system, none of these noise sources are as destructive as the noise caused by motion artefacts. This type of noise can be very destructive due to the change of distance between the sensing area of the electrode and the skin surface. This happens because the motion of the electrode in relation to the patient skin produces a voltage variation with higher value than the signal produced by the heart, leading to electronic saturation of analogue front-end equipment. If the ECG shape becomes distorted but not completely destroyed, typically it means that a noise with low frequency is overlapping the ECG signal. This problem are being addressed by different approaches based on digital signal processing techniques, as Empirical Mode Decomposition (EMD) [8] or Discrete Wavelet Transform (DWT) [1, 9]. Moreover, a solution based on accelerometers, used to measure the movements of the electrodes, can be used as correlated noise signal in adaptive filters [8, 10, 11, 14, 15]. However, all of these techniques work only if we have continuously signal information, *i.e.*, if the signal is not a voltage DC resulting from amplifiers saturation. In these cases, the solution should remove or predict the destroyed part of the ECG signal. For both cases the signal should be someway marked to inform the reader or user about the absence of real signal on that position. This paper focus in prediction the ECG signal using a linear prediction method based on Burg Algorithm.

3 Linear Prediction

Linear Prediction (LP) is a mathematical operation with ability to predict future values of a signal based on previous samples. The LP algorithms are based on frequency estimation; therefore, the results are as better as the knowledge about the number of signal sinusoids. This aspect is very important when we need to choose the algorithm to compute the LP coefficients. From the different techniques that can be used to compute the coefficients, the Burg algorithm [3] is the most used due to its stability [2]. Burg algorithm is based on Levinson-Durbin algorithm, where Burg slightly changed it to improve the stability.

3.1 Burg Algorithm

The Burg algorithm is based on forward LP:

$$f(n) = \widehat{s}(n) = -\sum_{k=1}^{m} \alpha_m(k)s(n-k), \tag{1}$$

and backward LP:

$$b(n) = \widehat{s}(n-m) = -\sum_{k=1}^{m} \beta_m(k)s(n+k-m). \tag{2}$$

Where $\hat{s}(n)$ is the prediction signal based on the previous m samples of the original signal $s(n)$ with N samples and $N > m$, finally α and β are the respective prediction coefficients. The Burg algorithm does not compute the coefficients directly; instead, performs the calculation of so called reflection coefficients ζ:

$$\zeta = \frac{-2\sum_{k=m+1}^{N} f(k)b(k-1)}{\sum_{k=m+1}^{N}\left(\left[f(k)\right]^2 + [b(k)]^2\right)}. \tag{3}$$

To minimize the sum of the squares error:

$$E = \sum_{k=m+1}^{N}\left(\left[f(k)\right]^2 + [b(k)]^2\right). \tag{4}$$

The algorithm recursively selects the ζ that minimises the average f and b error power.

4 System Architecture

The software implementation was divided in four main blocks as shown in Fig. 2. The ECG Noise Filter block is used for signal denoising, running an adaptive signal processing technique based on LMS algorithm as discussed in [12]. The buffer block is a simple buffer for ECG data samples. The buffer is useful to supply the prediction algorithm and error detection algorithm. The Burg algorithm block is used for estimation and run in parallel with buffering block; in fact the Burg algorithm uses the buffer content to estimate the next sample. The output samples from these two blocks are: $s(0)$ from buffer block and $\hat{s}(0)$ from Burg algorithm block. The $\hat{s}(0)$ sample is estimated with sample values from $s(1)$ to $s(m)$, where m is the size of the

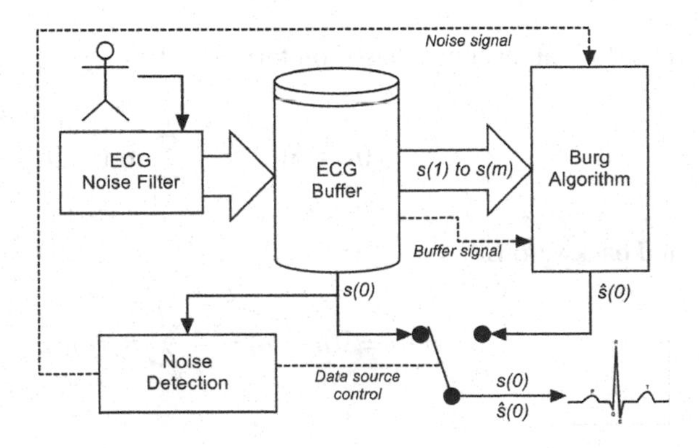

Fig. 2 Implementation block diagram

buffer. The buffer output is also the input of noise detection block, where if noise is present in the signal the output is switched from buffer output to prediction output. Noise detection block is based on the derivative of the signal and signal amplitude. If the signal suffers a fast variation, the derivative of the signal will be higher than the average value from previous ECG samples, which can be used as trigger for noise detection. Moreover, if the derivative of the signal is zero and the signal amplitude is equal to saturation value, we also are in the presence of noise. With the combination of these two techniques associated to the contents of the buffer, it is possible to detect the onset of noise artefacts as its length. There are also two different signals in block diagram, the buffer signal and the noise signal. These signals are control signals used on software to control the state of the program, where the buffer signal is used to inform about the availability of valid data in the buffer, and the noise signal is used to inform the algorithm block to not consider the noise present in the buffer.

5 Results Analysis

In order realise the efficiency of Burg Algorithm to predict the ECG signal, it was compared the results considering four different scenarios. In the first and second scenarios, the ECG signal was corrupted with motion artefacts, first with signal saturation and second with noise pulses with saturation. Next, in third scenario, it was evaluated the impact of buffer size in the estimation results, and in the fourth scenario

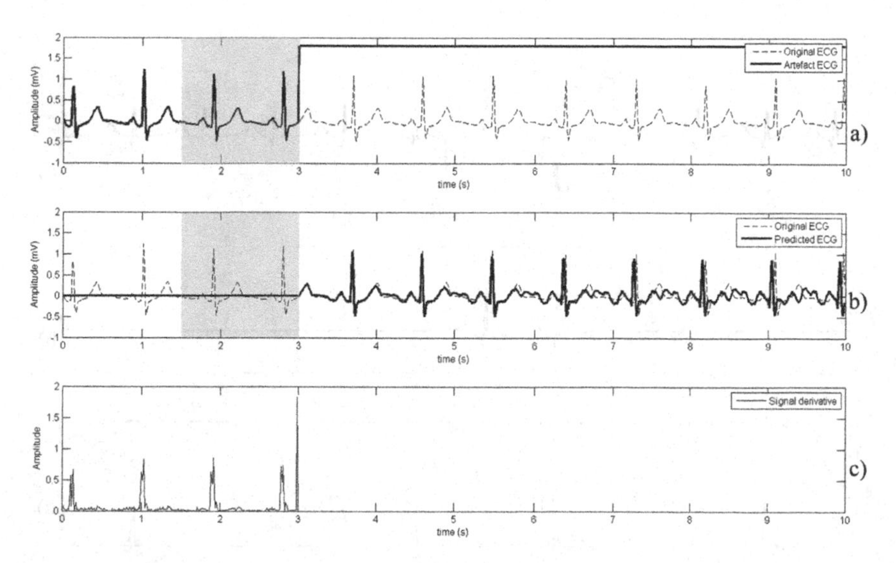

Fig. 3 Estimation result when signal is corrupted with signal saturation. **a** Signal saturation. **b** Estimation result. **c** Derivative

it was analyzed the impact of EMG noise in the algorithm. For algorithm implementation it was used the MATLAB tool and an ECG signal from MIT-BIH database. In the first experiment it was used signal saturation during several heart beats to realise the efficiency of the algorithm over time. The signal saturation is shown in Fig. 3a and the estimation result is shown in Fig. 3b. Figure 3c shows the derivative of original ECG signal. The size of the buffer is $m = 150$ which corresponds to almost two cycles of ECG been represented by the shadow area. It is possible to realise in Fig. 3b that in the first three ECG cycles the Burg algorithm is very efficient in signal estimation, were the differences between original (dashed line) and estimated signal (solid line) shows to be very small. Apart of a slightly difference in phase, the most relevant values are associated to the QRS complex, where the algorithm fails in the amplitude of R pike. Nevertheless, after the first three beats the signal shows several inconsistencies outside of QRG complex.
In the second experiment, the signal was corrupted with different noise signature, similar to pulses or spikes identical to the ones represented in Fig. 4a, the algorithm also responds very well as depicted in Fig. 4b. For this case, the noise was restricted to two ECG cycles and it is possible to see in Fig. 4c the difference of the derivative value at the onset and the end of the artefact.
In the third experiment, it was changed the value of m to realise the impact of the buffer size in the estimation. As expected, the estimated value is sensible to the minimum number of previous values used in estimation algorithm; therefore, the estimated values only make sense after minimum samples of signal. The minimum number m of samples is directly related with sampling frequency, *i.e.*, if the sampling frequency f_s of the AD converter increases, the m value also must increase.

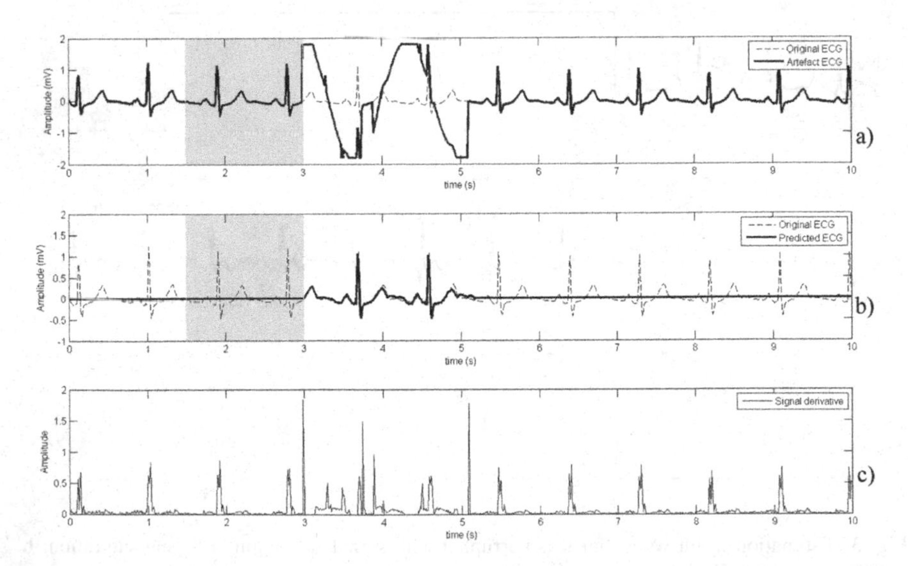

Fig. 4 Estimation result when signal is corrupted with noisy spikes. **a** Signal saturation. **b** Estimation result. **c** Derivative

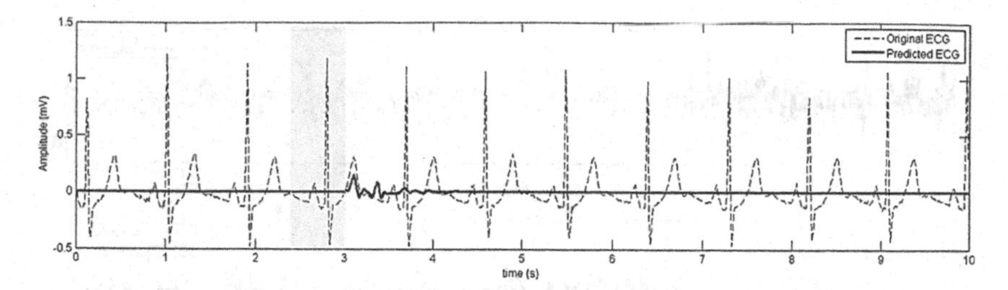

Fig. 5 Estimation result for buffer length smaller than one ECG cycle, where the *dashed line* is the original ECG and the *solid line* the estimation

Figure 5 shows the estimated signal when the length of the buffer does not contemplate at least one ECG cycle, where the prediction only happens for a short of samples and it is not capable to predict the entire cycle. However, if the buffer accommodate at least one ECG cycle, the algorithm can estimate several ECG cycles as shown in Fig. 6. In practice, the *m* value should be higher enough to accommodate at least one complete cycle of ECG signal. It is also important to consider the heart rate variability, therefore, the minimum number for *m* should be calculate for the minimum heart rate of human heart, typically between 60 and 80 bpm, and the f_s of AD. For that reason, the Burg algorithm block from Fig. 2 only starts outputting data after buffer signal is active, *i.e.*, after buffer is fulfilled with valid ECG data. An important aspect about the LP process is its susceptibility to fail in the presence of noise; therefore, in the fourth experiment it was used an ECG signal contaminated with random noise identical to EMG. The results are shown in Fig. 7b, where they are clearly worst when compared with filtered ECG. If the noise is similar to a sine-wave, as power lines noise, the algorithm will also reproduces the noise in estimation results, but if the noise is random, as EMG, the results will be random as well. In this work, it was used an adaptive algorithm for noise removal before signal buffering to remove this type of noise [12].

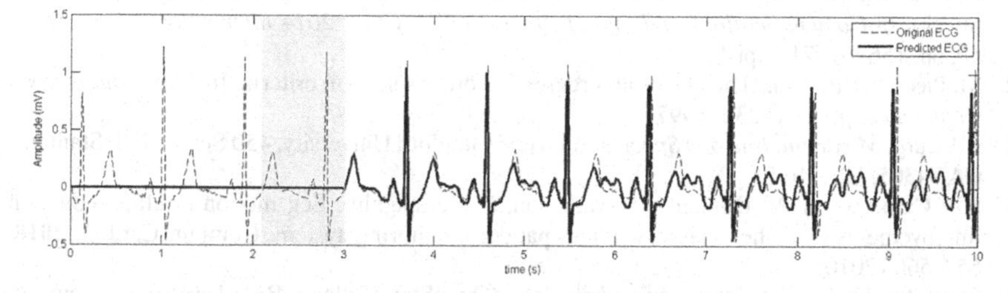

Fig. 6 Estimation result for buffer length equal to one ECG cycle, where the *dashed line* is the original ECG and the *solid line* the estimation

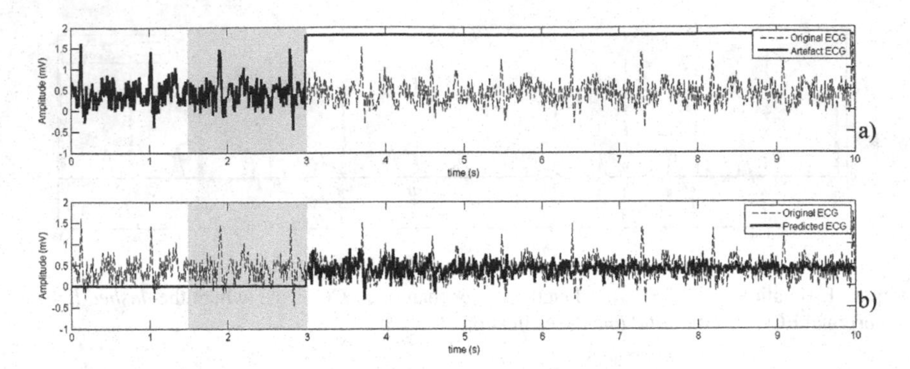

Fig. 7 Estimation result in the presence of random noise, where the *dashed line* is the original ECG with noise and the *solid line* the estimation

6 Conclusion

This work shows the ability of Burg algorithm to predict the ECG signal when noise produced by motion artefacts is destructive. The results are very promising since that, for the first three heart beats predicted, the results shown to be very similar to the original signal. It is also addressed the importance of the initial filtering of ECG signal, where in the presence of random noise similar to EMG noise the algorithm clearly fails. A final aspect respects to the computational requirements for portable devices, where the Burg algorithm also is very interesting. The computational activity is mainly multiplications and accumulations, which is well done by almost of the DSP processors in the market.

References

1. D.V. Bhoraniya, R.K. Kher, Motion artifacts extraction using dwt from ambulatory ecg (a-ecg), In *Communications and Signal Processing (ICCSP), 2014 International Conference on*, pp. 1567–1571, April 2014
2. M. Piet , T. Broersen, The abc of autoregressive order selection criteria, In *Proceedings Sysid Conference*, pp. 231–236, 1997
3. J.P. Burg, *Maximum Entropy Spectral Analysis* (Stanford University, 450 Serra Mall, Stanford, CA 94305, US, 1975)
4. P.A. Catherwood, N. Donnelly, J. Anderson, J. McLaughlin, Ecg motion artefact reduction improvements of a chest-based wireless patient monitoring system. Comput. Cardiol. **2010**, 557–560 (2010)
5. OpenStax College, *Anatomy and Physiology* (OpenStax College, Rice University, Houston, 2013)
6. A.C. Guyton, J.E. Hall, *Textbook of Medical Physiology*, 11th edn. (Elsevier, Philadelphia, 2006), pp. 19103–2899
7. D.-U. Jeong, S.-J. Kim, Development of a technique for cancelling motion artifact in ambulatory ecg monitoring system, In *Convergence and Hybrid Information Technology, 2008. ICCIT '08. Third International Conference on*, vol. 1, pp. 954–961, 2008

8. H.-K. Jung, D.-U. Jeong, Development of wearable ecg measurement system using emd for motion artifact removal, In *Computing and Convergence Technology (ICCCT), 2012 7th International Conference on*, pp. 299–304, 2012
9. M. Kirst, B. Glauner, J. Ottenbacher, Using dwt for ecg motion artifact reduction with noise-correlating signals, In *Engineering in Medicine and Biology Society, EMBC, 2011 Annual International Conference of the IEEE*, pp. 4804–4807, 2011
10. Y. Kishimoto, Y. Kutsuna, K. Oguri. Detecting motion artifact ecg noise during sleeping by means of a tri-axis accelerometer, In *Engineering in Medicine and Biology Society, 2007. EMBS 2007. 29th Annual International Conference of the IEEE*, pp. 2669–2672, 2007
11. S.H. Liu, Motion artifact reduction in electrocardiogram using adaptive filter. J. Med. Biol. Eng. **31**, 64–72 (2011)
12. A. Meireles, L. Figueiredo, L.S. Lopes, A. Almeida, Ecg denoising with lms adaptive filter, in *Proceedings of the 1st Conference on Electronics, Telecommunications and Computers on*, 2011
13. World Health Organization, *World Health Statistics* (2009)
14. M.A.D. Raya, L.G. Sison, Adaptive noise cancelling of motion artifact in stress ecg signals using accelerometer, in *Engineering in Medicine and Biology, 2002. 24th Annual Conference and the Annual Fall Meeting of the Biomedical Engineering Society EMBS/BMES Conference, 2002. Proceedings of the Second Joint*, vol. 2, pp. 1756–1757, 2002
15. D.A. Tong, K.A. Bartels, K.S. Honeyager, Adaptive reduction of motion artifact in the electrocardiogram, in *Engineering in Medicine and Biology, 2002. 24th Annual Conference and the Annual Fall Meeting of the Biomedical Engineering Society EMBS/BMES Conference, 2002. Proceedings of the Second Joint*, vol. 2, pp. 1403–1404, 2002

A Sentiment Analysis Classification Approach to Assess the Emotional Content of Photographs

David Griol and José Manuel Molina

Abstract The integration of Ambient Intelligence and Sentiment Analysis provides mutual benefits. On the one hand, Sentiment Analysis may enable developing interfaces providing a more natural interaction with human-computer interfaces. On the other, AmI enables using context-awareness information to enhance the performance of the system, achieving a more efficient and proactive human-machine communication that can be dynamically adapted to the user's state and the status of the environment. In this paper, we describe a novel Sentiment Analysis approach combining a lexicon-based model for specifying the set of emotions and a statistical methodology to identify the most relevant topics in the document that are the targets of the sentiments. Our proposal also includes an heuristic learning method that allows improving the initial knowledge considering the users' feedback. We have integrated the proposed Sentiment Analysis approach into an Android-based mobile App that automatically assigns sentiments to pictures taking into account the description provided by the users.

Keywords Human-computer interaction · Sentiment analysis · Emotion recognition · Mobile applications · Android

1 Introduction

Sentiment Analysis (SA), also known as Opinion Mining, has the main objective of extracting subjective emotional information from a natural language source [1–3]. Basic Sentiment Analysis algorithms are focused on classifying the input

D. Griol (✉) · J.M. Molina
Computer Science Department, Carlos III University of Madrid,
Avda. de la Universidad, 30, 28911 Leganés, (Spain)
e-mail: david.griol@uc3m.es

J.M. Molina
e-mail: josemanuel.molina@uc3m.es

A. Mohamed et al. (eds.), *Ambient Intelligence - Software and Applications*,
Advances in Intelligent Systems and Computing 376,
DOI 10.1007/978-3-319-19695-4_11

according to its polarity towards a specific topic (positive, negative, or neutral) [4]. There are also advanced approaches that add an additional level of granularity by further identifying private states, such as opinions, emotions, sentiments, evaluations, beliefs, or speculations [5].

Main applications of this field of study are currently related to marketing and social networks. Marketing applications are focused on determining customers' attitude towards products, which provides a very valuable information for companies to estimate products acceptance and market trends, offer products adapted to customers' requirements, and focus innovation on most demanded features.

The growing importance of Sentiment Analysis coincides with the growth of social media for sharing thoughts about trending topics in reviews, forum discussions, blogs, micro-blogs, Twitter, and social networks [6]. This allows analyzing the huge volume of information generated every day on the social media.

Main SA approaches in this application domain are lexicon or dictionary-based and machine learning methodologies. Lexicon-based models use a predefined set of words, which have an associated polarity. Document polarity will depend on the feature selection method and the combination of their scores. Machine-learning approaches usually rely on supervised classifiers. Although classifiers offer adaptability for specific contexts, they need to be trained with huge amounts of labeled data which may not be available, specially for upcoming topics.

In this paper, we describe a new proposal for Sentiment Analysis aimed to identifying the sentiments in the user's input, instead of only detecting the positive or negative polarity. The set of defined sentiments has been extracted from Plutchik's wheel of emotions [7], which defines eight basic bipolar sentiments and another eight advanced emotions composed of two basic ones.

Our proposal for SA combines a lexicon-based model for specifying the relationships between terms and emotions, and a statistical approach to identify the most relevant topics in the document that contribute the different sentiments. With this combination, we overcome the disadvantages of simple Bag-of-words models [4], which do not differentiate between parts of speech (POS). Bag-of-words models also weight all words commonly using the TF-IDF scheme (terms and their frequency), which usually leads to overweight most frequently used words. Furthermore, our proposal includes an heuristic learning method that allows improving the initial knowledge in the model by considering the users' feedback.

We have used our proposal for SA to develop an Android-based mobile application for emotion detection from photographs, which has been designed based on important studies about key aspects of algorithmic inferencing of emotions that natural images arouse in people [8, 9]. The *Emopic* App acts as a social network in which the users can share their photographs, know the emotions assigned to their descriptions, and compare how people from all around the world express their emotions about them.

2 Our Approach for Sentiment Analysis

The proposed model for Sentiment Analysis aims to extend common sentiment classification of text, which is usually focused on polarity, to a higher level so that the input texts are categorized by the emotions they evoke. Thus, the main goal is to recognize a specific set of human emotions instead of only detecting whether a piece of text is negative, neutral or positive. To do this, a limited set of emotions must be selected from one of the existing emotion classifications accepted by psychologist community.

After a detailed study of the principal affective models and considering computational requirements, we have selected a modification of the Hourglass emotion representation [10]. This model is based on Plutchik's wheel of emotions, which proposes eight basic emotions contrary to Ekman's initial classification that defines only six primary affection states. Although having more categories increases analysis complexity, Plutchik's model can be reduced into four categories -as there are four pairs of opposite emotions- so that, indeed, the analysis can be considered to turn out simpler. The proposed model is based on four key components.

The Knowledge Base (KB) contains the main information sources used by the Analysis Module to extract sentiment values from words. The Analysis Module completes the words analysis. By splitting texts in sentences an tokenizing words, this module can query the Knowledge Base to extract emotional information or know whether words are modifiers or carry an associated negation. Moreover, this module identify entities in the input text and track the number of occurrences of each one of them in a similar way bag-of-words models do this using occurrences vectors.

Once the entities have been identified and words are annotated with values from the KB, the Scoring Module computes the overall relevance of the entities and assigns a weighting factor for each of the words carrying emotional information, which are also known as concepts. A weight for each of the four independent emotional categories is then computed to classify the input text.

The last stage of the model deals with knowledge learning. To do this, the Learning Module takes as input the provided analysis from users when they disagree with the results of the Sentiment Analysis, and computes a learning factor to modify sentiment values of involved concepts.

2.1 Knowledge Base

As previously described, the Knowledge Base contains the main information sources used by the Analysis Module to extract sentiment values from words. In our proposal, this information has been classified into the following categories:

- **Concepts**: A concept refers to the emotions associated to a specific pair of (*word–PoS*), where PoS (part of Speech) denotes the grammatical function of a word inside a predicate. Only the primitive form of a word is considered and the rest of derivative words take the same set of emotional values. The different categories of words are:

- **Nouns**: Only the singular form is considered, although they may have an irregular plural that could be harder to identify. Nouns containing prefixes and suffixes are the only exception to this rule.
- **Adjectives**: The positive form is considered and both comparative and superlative forms are discarded.
- **Verbs** The infinitive form is considered. Some exceptions are made for -ing forms acting as a noun (e.g., "The professor's reading about macro-economics was brilliant')'.
- **Adverbs**: Only the positive form is considered, discarding comparative and superlative forms.

– **Modifiers**: Modifiers are denoted by an n-gram without associated sentiment states, which can increase, decrease or reverse the emotions of the associated concepts. They can be divided into two different categories:

- **Intensity modifiers**: This category is composed by those modifiers than may increase or decrease emotions expressed by concepts (e.g., "as much" or "a bit").
- **Negators**: These modifiers reverse the global emotion associated to a concept (e.g., "not" or "never").

The NRC[1] and SenticNet[2] emotion lexicons have been used to complete the KB. Both are publicly available semantic resources for concept-level Sentiment Analysis. A total of 12,297 concepts are currently stored in the KB.

2.2 Parser Module

The parsing process of a sentence generates its semantically analysis containing part-of-speech tags organized in a tree of predicates. Between the set of general-purpose libraries currently available, we have selected OpenNLP.[3] This library supports the most common NLP tasks, such as tokenization, sentence segmentation, part-of-speech tagging, named entity extraction, chunking, parsing, and coreference resolution.

OpenNLP uses the Penn Treebank notation,[4] which consider 36 sort of part of speech defined on the basis of their syntactic distribution rather than their semantic function. As a consequence nouns used in the function of modifiers are tagged as

[1] http://www.saifmohammad.com/WebPages/lexicons.html.

[2] http://sentic.net/.

[3] https://opennlp.apache.org/.

[4] http://www.cis.upenn.edu/~treebank/.

nouns instead of adjectives. Before parsing a text, it should be split into sentences by using the OpenNLP probabilistic *Sentence Detector*, which offers a precision of 94 % and a 90 % recall.

2.3 Emotion Classification Model

As stated before, our proposal uses an emotion representation model based on a modified version of the Hourglass model. The four independent categories that are considered for Sentiment Analysis consists of the following possible labels, described from negative maximum to positive maximum intensities, left to right:

- **Sensitivity**: [terror, fear, apprehension, neutral, annoyance, anger, rage]
- **Aptitude**: [amazement, surprise, distraction, neutral, interest, anticipation, vigilance]
- **Attention**: [grief, sadness, pensiveness, neutral, serenity, joy, ecstasy]
- **Pleasantness**: [loathing, disgust, boredom, neutral, acceptance, trust, admiration]

2.4 Text Scoring Scheme and Adaptive Learning

Once the parsing process has finished and all the concepts, modifiers and negators have been properly tagged, it is possible to begin with the computation of the sentiment values of the text. The scoring process follows a bottom-up approach based on a fixed algorithm that relies on the Knowledge Base accuracy, a proximity based approach for modifiers, and a topic detection module to detect the most relevant topics of a text.

The way sentences are weighted is based on entities occurrences. Let w_i be the weight of a predicate and n the total number of sibling predicates that are being combined, the sentiment value of a category for weighted predicates can be defined as:

$$S_w = \frac{\sum_{i=0}^{n} w_i * s_i}{\sum_{i=0}^{n} w_i}, \quad \begin{array}{r} \forall w_i > 0 \\ \forall s_i \neq 0 \\ s_i \in [-1, +1] \\ i = [0, n] \end{array} \tag{1}$$

Our proposal also integrates an adaptive learning process for improving the Knowledge Base used for Sentiment Analysis. This process uses Eq. 2 to consider the difference between the Sentiment Analysis output proposed by the SA algorithm and the feedback provided by the user. Let U be the set of sentiments of a text corrected by the user, M be the sentiments calculated by the SA algorithm, W_{C_s} be the weight of concept C for sentiment s, and A_c be the number of accumulated adjustments of concept C. Therefore the new value of each sentiment s for a concept C is defined as:

$$C_s = C_s + \frac{(U_s - M_s) * W_{C_s}}{1 + (A_C/1000)} \tag{2}$$

3 *Emopic*: Mobile Application to Assess the Emotional Content of Photographs

The visual component of our proposal is an Android-based mobile application for Android OS consisting of a social network for sharing photographs. The minimum Android version required to run the app is Android 4.1 Jelly Bean, which is currently supported by more than 70 % of mobile phones.

Regarding the main use cases of the application, we can distinguish three main processes: accessing the App, browsing the *Emopic* gallery, and posting images. Accessing the application requires registering and using the login information stored in the mobile phone.

Browsing images can be done through two different screens. The initial screen shows all pictures shared in the system in chronological order starting from the newest one. A second screen allows users to filter the *Emopic* gallery by the sentiments identified after the analysis of their description (Fig. 1-first).

The main functionality of the application is related to the posting process. This process includes the steps since the user decides to capture a photograph and share it on the App, to the assignation of sentiments considering the provided image description. The process can be divided in four stages: take photograph, type description, sentiment detection, and sentiment correction.

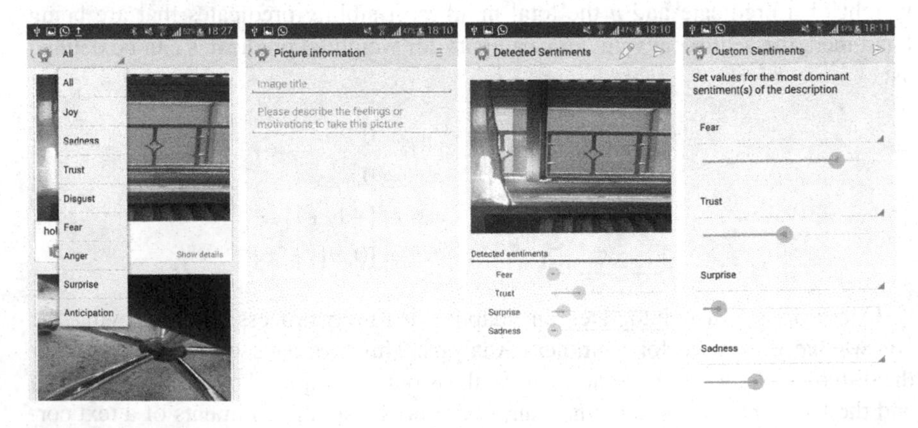

Fig. 1 Different screens of the *Emopic* mobile application

The first stage makes use of the default camera service client provided by Android. In the description screen (Fig. 1-second), the user provides a text describing the feelings or situations leading to the photograph which is about to be posted in a external server. Once the image is posted, the *Emopic* server starts running the implemented SA algorithm over the provided description. Figure 1-third shows the screen with the results of the analysis. As can be observed, the sentiments detected in the text are shown with their respective intensities (anger versus fear, sadness versus joy, disgust versus trust, and surprise versus anticipation.

To finish the posting process, users have two possibilities: either accept the results of the text analysis if they match user's real emotions or correct if the analysis is not accurate enough. If user chooses to click on accept button the post is finished and user is driven to main screen. However, in case the user's choice correspond to correcting the results, the screen shown in Fig. 1-fourth allows to select the right intensities for the sentiments expressed in the post description. After that, user just has to tap over the accept button to send the new values to the server so that the previous assigned intensities are substituted by the corrected ones and the post is updated.

The choice made by the user in this stage of the posting process has a significant meaning for the learning of the developed algorithm. On the one hand, whenever the user accept the results generated by the text analysis means that the text has been correctly classified and thus, the SA has succeeded in detecting the description's emotions. On the other hand, every time a user disagree with the outcomes of the analysis and chooses to provide the right sentiments of the introduced description, the new sentiments are not only used to update the classification, but also to update the algorithm by adjusting concepts weights as explained in the previous section.

From the design point of view, one the main characteristic of *Emopic* is that it acts as a social network, in which all posted photographs are publicly available. By publicly sharing all the uploaded pictures, the *Emopic* gallery will be an opportunity for comparing how people from all around the world express their emotions through their photographs.

A preliminary evaluation of the application has been already completed with the participation of 33 recruited users. The questionnaire with the following questions was defined for the evaluation: *Q1*: On a scale from 1 to 5, how much experience with smartphones do you have?; *Q2*: On a scale from 1 to 5, how much experience with Android do you have?; *Q3*: How often do you use image-sharing social networks (e.g., Instagram)?; *Q4*: On a scale from 1 to 5, how much do you know about Sentiment Analysis?; *Q5*: On a scale from 1 to 5, how understandable the steps required in the different functionalities of the App are?; *Q6*: On a scale from 1 to 5, how accurate the results of the Sentiment Analysis are?; *Q7*: Was it easy to use the App? The users were also asked to rate the system from 0 (minimum) to 10 (maximum) and there was an additional open question to write comments or remarks.

The results of the questionnaire are summarized in Table 1. As can be observed from the responses to the questionnaire, most of the users participating in the survey use smartphones and the Android OS, and not all of them usually access image-sharing social networks. Few of them had a previous knowledge about Sentiment Analysis.

Despite most of the participants agree with the set of sentiments used for representing emotions, almost half of them would prefer to have a larger list of sentiments available in the tool. Regarding accuracy, most of the users agreed that the overall performance of the analysis of their texts were from 3 to 5. Most of the users found the application easy to use. The satisfaction with technical aspects of the application was also high, as well as the perceived potential to recommend its use. The global rate for the system was 8.6 (in the scale from 0 to 10).

Table 1 Results of the evaluation of the *Emopic* App by recruited users

	Min / max	Average	Std. deviation
Q1	4/5	4.17	1.19
Q2	3/5	3.06	1.47
Q3	2/5	2.83	1.25
Q4	1/3	2.05	0.70
Q5	4/5	4.17	0.68
Q6	3/5	4.63	0.45
Q7	4/5	4.83	0.37

4 Conclusions

In this paper, we have presented a novel Sentiment Analysis approach that combines a lexicon-based model for specifying the set of emotions and a statistical methodology to identify the most relevant topics in the document that are the targets of the sentiments. Our proposal for SA also includes an heuristic learning method that allows improving the initial knowledge considering the users' feedback. By means of our proposal, we overcome the main disadvantages of Bag-of-words models, which do not differentiate between parts of speech and usually lead to overweight most frequently used words. In addition, our proposal includes an heuristic learning method that allows improving the algorithm by updating the Knowledge Base.

We have used the proposed Sentiment Analysis approach to develop an Android-based mobile App that automatically assigns sentiments to pictures taking into account the description provided by the users. As future work, we want to extend the preliminary evaluation of the application to improve the proposed SA algorithm and carry out a comparative assessment with other SA algorithms. With the results of these activities, we will optimize the system, and make the application available in Google Play.

Acknowledgements This work was supported in part by Projects MINECO TEC2012-37832-C02-01, CICYT TEC2011-28626-C02-02, CAM CONTEXTS (S2009/TIC-1485).

References

1. B. Liu, *Sentiment Analysis and Opinion Mining. A Comprehensive Introduction and Survey.* (Morgan & Claypool Publishers, San Rafael, 2012)
2. W. Medhat, A. Hassan, H. Korashy, Sentiment analysis algorithms and applications: a survey. Ain Shams Eng. J. **5**(4), 1093–1113 (2014)
3. W. Medhat, A. Hassan, H. Korashy, Computational approaches to subjectivity and sentiment analysis: present and envisaged methods and applications. Ain Shams Eng. J. **5**(4), 1093–1113 (2014)
4. B. Pang, L. Lee, S. Vaithyanathan, Thumbs up?: sentiment classification using machine learning techniques, in *Proceedings of the CEMNLP* (2002), pp. 79–86
5. J. Wiebe, E. Riloff, Creating subjective and objective sentence classifiers from unannotated texts. LNCS **3406**, 486–497 (2005)
6. F. Pla, L. Hurtado, Political tendency identification in twitter using sentiment analysis techniques, in *Proceedings of the COLING* (2014), pp. 183–192
7. R. Plutchik, H. Kellerman, *Emotion: Theory, Research and Experience*, vol. 1, Theories of Emotion. (Academic Press, San Diego, 1980)
8. Datta, R., Li, J., Wang, J.: Algorithmic inferencing of aesthetics and emotion in natural images: an exposition, in *Proceedings of the ICIP* (2008), pp. 105–108
9. C.H. Chen, M.F. Weng, S.K. Jeng, Y.Y. Chuang, Emotion-based music visualization using photos, in *Proceedings of the MMM* (2008), pp. 358–368
10. E. Cambria, A. Livingstone, A. Hussain, The Hourglass of Emotions. LNCS Cognitive Behav. Syst. **7403**, 144–157 (2012)

Opportunistic Sensoring Using Mobiles for Tracking Users in Ambient Intelligence

Javier Jiménez Alemán, Nayat Sanchez-Pi and Ana Cristina Bicharra Garcia

Abstract The necessity of using new technologies to monitoring elderly people in open-air environments by their caregivers has become a priority in the last years. In this direction, Ambient Intelligence (AmI) provides useful mechanisms and the geo-localization technologies embedded in smartphones allows tracking elderly people through opportunistic sensoring. The aim of this paper is to show a practical example to how to combine some technologies for monitoring elderly people through the system *SafeRoute*. We describe the two components of this system: the Android application *CareofMe* and the web system *SafeRoute*. The proposed system uses GPS, Wifi and accelerometer sensoring, GoogleMaps functionalities in Android and web environments and an alert system for caregivers.

Keywords Opportunistic sensoring · Ambient intelligence · Elderly tracking · Fall detection · Geo-localization technologies

J.J. Alemán (✉)
Institute of Computing, IC. Fluminense Federal University, Rua Passo da Pátria, São Domingos, Niterói, Rio de Janeiro, Brazil
e-mail: jjimenezaleman@ic.uff.br

N. Sanchez-Pi · A.C.B. Garcia
ADDLabs. Documentation Active and Intelligent Design Laboratory of Institute of Computing, Fluminense Federal University, Av. Gal.Milton Tavares de Souza. Boa Viagem, Niterói, Rio de Janeiro, Brazil
e-mail: nayat@addlabs.uff.br

A.C.B. Garcia
e-mail: cristina@addlabs.uff.br

A. Mohamed et al. (eds.), *Ambient Intelligence - Software and Applications*,
Advances in Intelligent Systems and Computing 376,
DOI 10.1007/978-3-319-19695-4_12

1 Introduction

In recent decades, there has been a global trend of increasing average age of the population and life expectancy. According to the statistics of the census of the Brazilian Institute of Geography and Statistics (IBGE) in 2010, the Brazilian population over 65 years old increased in a 10.3 %, this percentage will reach 29 % in 2050 and 36.1 % in 2075 [1]. In [2], authors estimate 1 million 200 thousands people living with Alzheimer disease in Brazil, with 70 % living in their own homes. This fact implies an increase of the permanent attention to these people by caregivers and an increasing demand of support mechanisms for these tasks.

The European Community's Information Society Technology (ISTAG), belonging to the European Commission, defined the concept Ambient Intelligence (AmI) in 2001. Ambient Intelligence is an emergent topic that attempts to answer human needs through digital and technological environments, allowing innovative ways of human-computer interactions [3]. The Ambient Intelligence based technologies for the support to daily activities are also called tools of Ambient Assisted Living (AAL). AAL can be used in prevention of accidents and to improve the health conditions and comfort of the elderly people [4]. These technologies can supply security to the elderly, developing response systems for mobile systems, fall detection systems and video surveillance systems [5].

Nowadays, most of the smartphones not only serve as communication devices, but also are equipped with several sensors like accelerometer, gyroscope, proximity sensors, microphones, GPS system and camera. All these sensors make possible a wide range of applications like the assistance to people with disabilities, intelligently detecting and recognizing the context.

The unobtrusive sensoring is also called opportunistic sensoring and represents a desirable characteristic of the AAL systems. There are several initiatives to develop techniques of opportunistic sensoring in mobile devices in the last years, allowing the creation of a new kind of mobile applications in the context of the Ambient Intelligence for the care of elderly people.

To fulfil the needs of tracking, current mobile phones are equipped with several positioning methods that are based on the Global Positioning System (GPS), WiFi or Cell-Id, which mostly results in a high-energy demand and thus quickly drain the device's battery [6]. While GPS allows for an accuracy location up to 5–10 m, it requires several seconds to minutes to determine the position. In addition, the functionality is limited, for instance inside buildings. Wifi based location is sufficiently accurate with 30–50 m. However, it required the availability of a registered wireless hotspot, which may only be found in densely populated areas. Cell based location is available as long as there is mobile network coverage. The downside of this technology is that it is less accurate with a deviation of several hundreds of meters [7]

Fall detection is an important component of many AAL systems because approximately half of the hospitalizations of the elderly are caused by fall [8]. Not only are fall related injuries the number one reason for emergency room visits, it is also the leading cause of injury-related deaths among adults 65 years old and older [9].

Taking into account all these considerations, we conclude that elderly people often suffer problems of aging such as memory loss, difficulty walking, etc. Many times, these people have to stay at home alone for long periods, but they normally do various activities outside the home during this time (go to the market, visiting friends, etc.). Once they are in outdoor environments, elderly people are at risk of fall down or getting lost on the way. In these cases, it is very important that elderly people can communicate with their carivegers for help and receive orientations in real time.

The purpose of this work was to presents the system *SafeRoute*, a system able to assist elderly people with activities related to their day-to-day activities in open-air environments, and using the geo-localization technologies inside of mobiles devices. We attempt to develop a system that opportunistically monitored elderly people who follow predefined routes and efficiently notified their caregiver in case of emergency.

The paper is organized as follows: Sect. 2 presents the analysis of some studies about the use of mobiles as AAL tools for the monitoring of elderly people in open-air environments. Section 3 describes the design and the implementation of *SafeRoute* with its contributions and limitations. Finally, in Sect. 4, some conclusions are given and future improvements are proposed.

2 Related Work

The availability of smart phones equipped with a rich set of powerful sensors at low cost has allowed the ubiquitous human activity recognition on mobile platforms. There are several advantages in the use of smartphones, for example, the developments kits (SDK), APIs and mobile computing clouds allow developers to use backend servers and collect data from a large number of users.

Recent studies show that smartphones are suitable for tracking, monitoring the position of the elderly and alerting in case of estrangement from a predefined route. *SmartShoe* [10] is a system where the authors combine use of a device in the shoe of the elderly person, equipped with technologies GPS and Bluetooth and a web interface to allow caregivers to make an efficient tracking. *Navitime* [11] is another system that helps pedestrian to find the best route to his destiny through different kinds of transport and attending parameters like the estimated delay time or the estimated amount of carbon dioxide that can be expelled. This system guides users through maps, predefined routes, voice alerts and vibration. *Navitime* accuracy is 10 m in unobstructed areas and 3 m in highly urbanized areas, in these cases using map matching methods and estimation techniques through cell towers positioning.

Several strategies for monitoring and tracking, and related types of interventions have been implemented with mobile phones: (1) tracking people trying to optimize the use of energy in the mobile device [12–14]; (2) the inclusion of social networks [15–17].

The authors of *EnLoc* [12] focused their work on the optimal use of energy when location sensors in smartphones are used. They developed a localization framework that is able to detect the user optimal localization and to predict the estimated time of the energy use through heuristical predictions.

LibreGeoSocial [15] is a very interesting study that combine healthcare and social networks. Authors created a mobile social network, which allows creating virtual communities to facilitate communication between elderly persons and their caregivers in case of loss. There are different scenarios, for example, carivegers can put a perimeter around a predefined route or at home for the elderly and the system will send an alarm and a message through the network to report the current position if distancing.

There are also studies that deal AAL modeling knowledge through ontologies as COMANTO [18], SOPRANO [19] and SSH [20]. On the other hand, there are studies that emphasize classification algorithms for reasoning about knowledge already represented: (1) tree decisions [15, 18]; (2) K-Means algorithm [21].

COMANTO is one of the examples mentioned above of modelling of knowledge using ontologies. COMANTO authors focused their attention on creating a generic ontology based in a localization context; they try to describe a general context of interrelations between concepts in a nonspecific situation in pervasive and distributed environments. Among the main classes in this ontology, the class "Person" is the central class. In addition, there are properties that make possible to incorporate the relations to the context like "friendOf" to represent the associations "Person" to "Person". There are other important classes like "Place"that represents the abstraction of a physic spatial place, "Activity", "Agenda", "Physical Object", "Sensor" and "Preferences. Finally, the class "Time" represents the crucial information related with the actual time, acting as timestamp for that context information that can change over time.

Different examples show how AAL systems provide useful information in real time. Many systems implement web services because of friendliness of the web interfaces for most users. The web services are very useful because the involved technologies (HTTP, XML) are independent of programming languages, platforms and operational systems. For example, [10] makes possible do the tracking of an old person's route through a web site that sends alerts in case of distancing. On the other hand, in [17], authors implement a solution that provides the exact position of the old person in Google Maps using a social network and shows the user location through a radar when the map information is lost or disabled.

In our approach, we create a system that combines functionalities of route monitoring and fall detection through sensors built-in smartphones to sends alerts to carivegers in case of emergency. System is composed by two main components of the system (the Android application *CareofMe* and the web service *SafeRoute*). These two components work in a combined way and merge information from sensors embedded in mobiles devices for tracking elderly people. It is also presented as future work, a group of challenges to implement in our system to improve the quality of life of older people in outdoor environments.

3 SafeRoute

The proposed system (Fig. 1) is composed by two components: the Android application *CareofMe* and the web system *SafeRoute*.

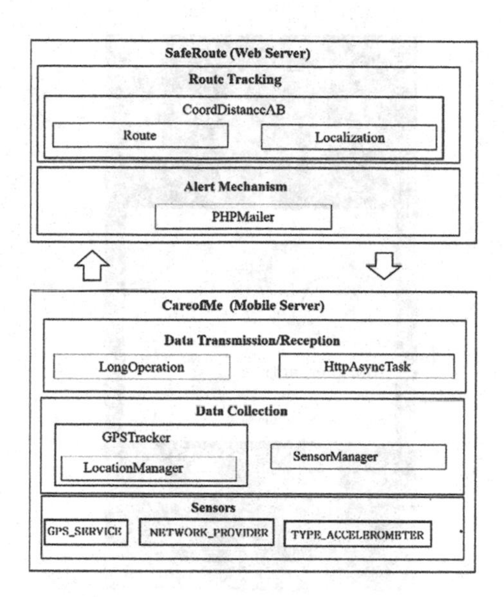

Fig. 1 System architecture

The *CareofMe* app works as follows; for both functionalities, firstly, the monitored user chooses a predetermined route to follow (Fig. 2). *CareofMe* application uses a combination of *GPS* and Wifi technologies to show the current user localization in an open-air environment. We decide to include the use of the Wifi service because is a service built-in inside smartphones that can be used like support operation in case of fails in the *GPS* service and increases the saving of energy. In the case of route monitoring (Fig. 3), *GPSTracker* is the responsible class for managing position dates through the class *LocationManager*, belonging to the API responsible for working with maps in mobile environments: GoogleMap Android v2. For fall detections, we used the accelerometer sensor to measure the coordinates (x, y, z) of the smartphone position and detect abrupt changes to indicate a fall. *SensorManager* is the Android class responsible to manipulate the values of the accelerometer sensor. In case of fall detection, the app will send the geo-localization coordinates to the web server if detect a fall and will ask to user if he need some kind of help (Fig. 4).

On the other hand, *SafeRoute* system was conceived as a web service for the constantly monitoring of the user's status and the sending of alerts to carivegers and the elderly person in case of distancing or fall. The Route Tracking functionality uses the *CoordDistanceAB* class to calculate distance between user locations received in real-time (*Localization*) and the locations of the predefined route (Route). *PHPMailer* sends an email to the cariveger with the exactly position of the elderly person in case of distancing.

Fig. 2 Predefined route selection

Fig. 3 Route monitoring

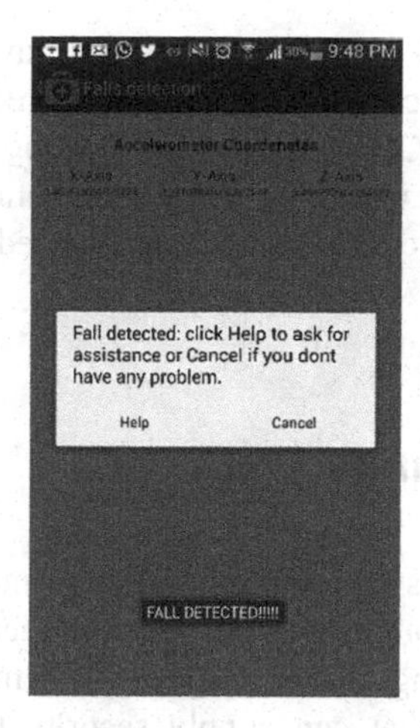

Fig. 4 Fall detection

On the other hand, it was conceived the web system *SafeRoute* (Fig. 5) as a web service for the constantly monitoring of the user's position and the sending of alerts to the old person in case of distancing or fall. The reasons for using web technology were some of the advantages mentioned above, for instance, the non-dependence of any programming language to access this type of system.

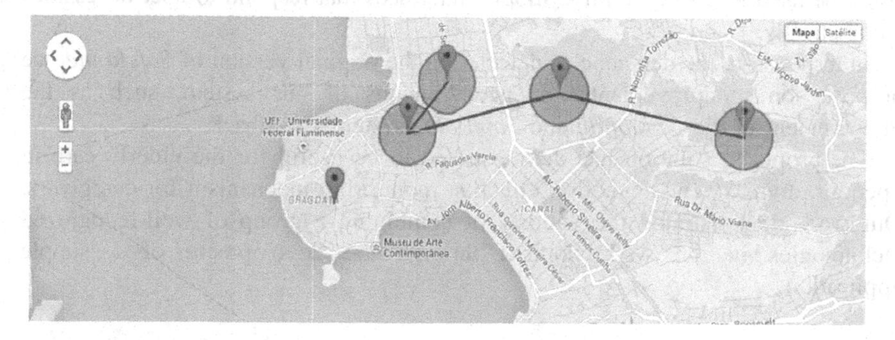

Fig. 5 *SafeRoute*

SafeRoute was developed using PHP 5.0 and Javascript, the predetermined routes of the users and the localization data were stored in a MySQL database. In addition, the system use the GoogleMap v3 and Google Maps JavaScript API v3 APIs. These APIs allow the use of a group of functionalities like the drawn of polylines for the routes and circles to indicate the radius of allowed distance of distancing in each point.

4 Conclusions and Future Work

The use of mobile devices is increasing gradually, making these devices a new source for developing solutions in various technologies. In AmI, AAL is gaining more prominence by providing mobile response systems, fall detection systems or video surveillance systems that can supply security to the elderly and to their caregivers.

The potentialities of the geo-localization technologies built-on in smartphones has been used in the last years for tracking elderly people in open-air environments.

In this paper, we presented the AAL system *SafeRoute*, wich combines two geo-locations based functionalities for the care of elderly people: route monitoring and fall detections.

We believe that our system can improve its operations in many aspects and we identified a group of future works. For instance, we think that mechanism of feedback proposed is considered poor because only reported to the old person about the distance of the predetermined route and not report to old people's caregivers about the position of the elderly person in case of distancing. We also find problems in the web interface of *SafeRoute* because it´s not enough intuitive considering all the potentialities of the web design (Example: The system could propose path to follow in case of distance of the old person). We also believe that it is possible using other sensors built-in smartphones to create new functionalities in the system, ambient temperature, relative humidity, light and proximity sensors could be used to create intuitive mobile interfaces that respond to user necessities automatically.

In response to the deficiencies detected in the original version of *SafeRoute*, we propose some improvements for new versions of our system, such as the improvement of the *CareofMe* and *SafeRoute* system's feedback.

The proposed solution has demonstrated to be useful for the elderly care in open-air environments, enabling effective monitoring mechanism for caregivers. Our work demonstrated the validity of combining a group of well-recognized technologies in the AAL context through the development of a simple application.

References

1. Instituto Brasileiro de Geografia e Estatística (IBGE). Sinopse dos Resultados do Censo (2010), http://www.censo2010.ibge.gov.br/sinopse/webservice/
2. J.L. Nealon, A. Moreno, *Applications of Software Agent Technology in the Health Care Domain* (Birkhauser Verlag AG Whiteistein Series in Software Agents Technologies, Bases, Germany, 2003)
3. F. Sadri, Ambient intelligence: a survey. ACM Comput. Surv. **43**(4), 36:1–36:66, Oct (2011), http://doi.acm.org/10.1145/1978802.1978815
4. P. Rashidi, A. Mihailidis, A survey on ambient-assisted living tools for older adults. IEEE J. Biomed. Health Inform. **17**(3) (2013)
5. M. Roussaki et al., Hybrid context modeling: a location-based scheme using ontologies. in *Proceedings of the Pervasive Computing* and *Communications Workshops* (2006)
6. U. Bareth, A. Kupper, Energy-efficient position tracking in proactive location-based services for smartphone environments. in *35th IEEE Annual Computer Software and Applications Conference* (2011)
7. S. von Watzdorf, F. Michahelles, Accuracy of positioning data on smartphones. in *3rd International Workshop on Location and the Web* (2010)
8. H. Gjoreski et al., RAReFall - Real-time activity recognition and fall detection system. in *IEEE International Conference on Pervasive Computing and Communications Demonstrations* (2014)
9. F. Sposaro, G. Tyson, iFall: An android application for fall monitoring and response. in *Annual International Conference of the IEEE* (2009)
10. B. Silva, J. Rodrigues, An ambient assisted living framework for mobile environments (2013)
11. M. Arikawa, S. Konomi, K. Ohnishi, *Navitime*: Supporting pedestrian navigation in the real world. IEEE Perv. Comput. **6**, 21–29 (2007)
12. I. Constandache et al., Enloc: energy-efficient localization for mobile phones, INFOCOM (2009)
13. M. Kjærgaard, J. Langdal, T. Godsk, T. Toftkjær, Entracked: energy-efficient robust position tracking for mobile devices. in *7th International Conference on Mobile Systems* (2009)
14. J. Paek et al., Energy-efficient rate-adaptive gps-based positioning for smartphones. in *8th International Conference on Mobile Systems, Applications, and Services* (2010)
15. E. Miluzzo, Sensing meets mobile social networks: the design, implementation and evaluation of the CenceMe application. in *6th ACM Conference Embedded Network Sensor Systems* (2008), pp. 337–350
16. S. Gaonkar et al., Micro-Blog: sharing and querying content through mobile phones and social participation. in *6th International Conference on Mobile Systems, Applications, and Services* (2008)
17. R. Calvo-Palomino, P. de las Heras-Quirós, Outdoors monitoring of elderly people assisted by compass, GPS and mobile social network (2009)
18. I. Roussaki et al., Hybrid context modeling: a location-based scheme using ontologies. in *Proceedings of the Pervasive Computing* and *Communications Workshops* (2006)
19. M. Klein, A. Schmidt, R. Lauer, Ontology-centred design of an ambient middleware for assisted living: the case of soprano. in *Proceedings of the Annual German Conference Artificial Intelligence* (2007)
20. L. Chen et al., Semantic smart homes: towards knowledge rich assisted living environments. in *Proceedings of the Intelligence Patents Management* (2009)
21. J. Yang, Toward physical activity diary: motion recognition using simple acceleration features with mobile phones. in *1st International Workshop on Interactive Multimedia for Consumer Electronics* (2009), pp. 1–10

Mobile Crowd Sensing for Solidarity Campaigns

Ana Alves and David Silva

Abstract We present an ongoing project (This work is partially supported by the InfoCrowds project-PTDC/ECM-TRA/1898/2012 This work is supported by CISUC, via national funding by the FCT - Fundação para a Ciência e Tecnologia.) which has two separate strands, one refers to the technological study about the applicability of high performance and high availability technologies in Web Services and the other is directed to a practical application of these technologies to solidarity campaigns in collecting goods. The focus of this paper is in the first one, a technological study where several frameworks for building Web Services, databases of different types and libraries to assist in obtaining product codes (barcodes) and data are analyzed, this includes a study of performance, availability and reliability, as well as appraisals for each one. Besides this, we introduce an experimental setup and results obtained so far in a third sector institution, Caritas Diocesana of Coimbra (http://www.caritas.pt/site/nacional/ Portuguese Website (last visited in March 2015)), a non-profit organization part of Caritas (http://www.caritas.eu/ (last visited in March 2015)). As main contribution, we propose a distributed architecture for Mobile Crowd Sensing able not only to allow real time inventory through simultaneous campaigns but also it gives feedback to volunteers in order to instantly acquire information about which categories of goods are more needed.

Keywords Mobile crowd sensing · Android · Nosql · Mobile applications · Solidarity campaigns

---

A. Alves (✉) · D. Silva
Polytechnic Institute of Coimbra, Coimbra, Portugal
e-mail: aalves@isec.pt

D. Silva
e-mail: a21170222@isec.pt

A. Alves
Centre of Informatics and Systems, University of Coimbra, Coimbra, Portugal

A. Mohamed et al. (eds.), *Ambient Intelligence - Software and Applications*,
Advances in Intelligent Systems and Computing 376,
DOI 10.1007/978-3-319-19695-4_13

1 Introduction

Crowd Sensing, also known as community sensing [1], participatory sensing [2] or opportunistic sensing [3], ranges from the active participation of users to contribute data (e.g. when taking a picture, report the closing of a road, reporting an accident) until an autonomous data collection using sensors and with a minimum involvement (e.g. continuous collection of location data without explicit user action). Crowd sensing is sometimes classified depending by which way data is collected: from mobile devices (Mobile Crowd Sensing) or Web (Web Crowd Sensing).

The typical scenario for Mobile Crowd Sensing is based on simple data collection from sensors included on mobile devices. Sometimes these data are locally processed to get useful information being sent to a service/server for aggregation and analysis.

Web Crowd Sensing is voluntary data collection when proprietary data acquisition becomes extremely expensive (e.g., points-of-interest introduced into Foursquare,[1] bar codes and descriptions of products populating a universal database universal[2]).

Due to the social and economic context in the World, several national solidarity campaigns have emerged [4], and for the first time a global campaign[6], in recent years to collect food and other goods (books, toiletries, school supplies, etc.). These campaigns may last for a short period of time and be promoted in the media supported by large commercial surfaces (e.g. Food Bank campaigns[3]) or may have a more permanent basis in gathering goods through Web sites and fixed collection points across a country (e.g. the localreuse project[4]).

Despite these campaigns are extremely helpful, they raise two problems:

- Discrepancy between need and supply - Although all help is welcome, there are more priority goods, priorities that can change based on prior campaigns, time of year, natural phenomena (e.g.: floods, fires, earthquakes) or even a function of the current campaign since there are themes of a particular type (e.g. feeding, toiletry, school supplies) but not specific to a specific good. In the latter case, it is advantageous for the volunteer (for both the giver as to who is collecting) know in real time which the amount of specific categories of goods have been donated so far (e.g. tons of rice, boxes of diapers, thousands of pencils).
- Geographic heterogeneity - Having people donating goods around a country, and taking into account the population dispersion, likely this donation is not homogeneous. Even in the regional level there is a great discrepancy in terms of

[1]http://www.foursquare.com (last visited in March 2015).

[2]http://upcdatabase.org/ (last visited in March 2015).

[3]News about worldwide campaigns (http://www.eurofoodbank.eu/portail/index.php?option=com_content&view=article&id=226 and http://food.caritas.org/pope-urges-catholics-to-join-caritas-week-of-action-to-end-hunger/ last visited in March 2015).

[4]http://localreuse.org/ (last visited in March 2015).

quantity and type of donations, something if it could be possible to estimate, would allow better optimization of human and logistical resources

Having been successfully applied to mobility data and urban planning [6, 7], this work aims to apply the concept of Mobile Crowd Sensing in real time to solidarity campaigns, a new field for a proof of concept. In addition to its Software Development side, this project will contribute to the inventory of goods, analysis of the data collected and innovates due the inherent social objective. Concretely, we proposed and implemented an easily adaptable platform[5] that incorporates the concept of Mobile Crowd Sensing. The purpose behind this platform is to ease and centralize the inventory's process, in real time and easy to use, in order to encourage people to contribute in balanced portions, with what is really needed in a higher priority. This led to a collection of applications optimized in accordance with the objectives established by the institution Caritas Diocesana of Coimbra, the host of this project. To achieve this goal, these applications use the latest technologies such as Android, Web Services and non-relational databases.

The remainder of this paper comprises 3 sections. Section 2 introduces the institution where this project is set up. Section 3 is devoted to the proposed architecture with each module described in detail. And finally, Sect. 4 presents some conclusions and future work.

2 Community Engagement

Leveraging urban technology for social purposes, such as volunteer participation and community engagement, has been steadily gaining interest in the scientific community and in the third sector. Essentially, citizens can be empowered to participate and engage the city better through the use of modern technology. Examples of these technologies are mobile phones, public displays, sensor networks, art installations, or any other type of urban technology. A national social study was made in USA [7] where results show that the diversity of mobile phone use and the frequency of relational mobile communication are positively associated with mobile donation. Internet donation and mobile donation complement rather than compete with each other.

Mobile Crowd Sensing until now, as far we know, never was used in this context to help a Solidarity Institution in the real time sensing of the adherence in campaigns for collecting goods. Besides the fact we are implementing this pilot in a regional coverage, we believe this project can be adopted by others NGO in a broader sense.

Caritas is an Institution of Social Solidarity supporting transversely communities in social, health, education and pastoral care. Implemented since the 50s, Caritas

[5]http://kenobi.dei.uc.pt:8080 where the user can download the Mobile App and interact with the Crowd Sensing Platform (last visited in March 2015).

has always tried to follow and respond to the problems of the communities, using a methodology that favors dialogue, cooperation and networking. Currently it focuses its intervention in the search for innovative and economically sustainable strategies, which enable the provision of answers with quality, suitable to emerging needs while maintaining focus as humanism, professionalism and technical and scientific rigor. With an increasing technology adoption, Caritas has been involved in the development of some European projects [8], being an enthusiastic collaborator in the proposed architecture. Thus, it will be possible to integrate in a near future the Mobile Crowd Sensing platform with the intranet which actually exists in that institution.

3 Distributed Architecture

One way to address the problem of the growing number of users and the need to provide more and more information efficiently passes precisely by how it is processed and made available. Exactly on these subjects, high performance and high availability, we built our platform, being as a requirement to serve a high number of users and great amount of information. Figure 1 presents two possible configurations of our system depending on the availability of resources: a minimal and a maximal topology. In the latter, geographically distant servers were tested and the database was also maintained in a distributed and synchronized parallelism.

This project was developed following an agile methodology, with cyclical definition phase of requirements, architecture design, development, testing and presentation of release's result, as a way to get feedback and allow a more efficient development of the project. Its main components are: a mobile application with the ability to read EAN[6] bar codes; a very responsive and high available Web service which serves as interface to a non-relational Database Management System (DBMS).

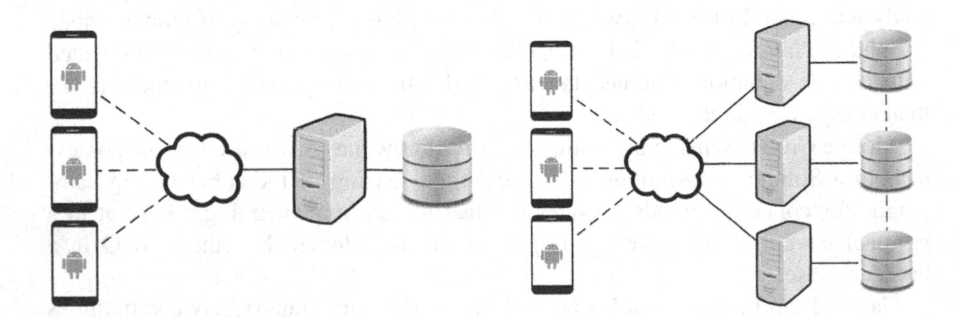

Fig. 1 Minimal e maximal topology of the proposed architecture

[6]International Article Number (originally European Article Number).

3.1 Web Service

High performance frameworks use a different strategy from common thread-based services, taking advantage of an event-driven architecture and a non-blocking API, usually coupled with the reactive software pattern. In practice this means to have a single thread that is cyclically waiting for new events, these may be a new application or the appointment of new requested data, thus reducing wait cycles due to hardware access delays, reducing memory use and sharing best processor time for each client. Priority was given to frameworks prepared for high availability, reliability and performance, preferably linked to successful cases and recommended by the respective communities for similar projects.

A highlight point is that most of these frameworks use a reactive pattern instead of the typical model of worker-process, thread creation or pool of threads. Thereby it maximizes the number of requests accepted without congestion or depletion of resources in process data access. In this line, based on the results of tests performed over various frameworks (ribs2,[7] Vert.x,[8] Netty,[9] Tomcat,[10] EventMachine,[11] Node.js[12]), it was decided to opt for Vert.x which presents good response time and a very low error rate, and additionally, it allows both server and client side being developed in the same language (Java).

The chosen solution was to create a Verticle that implements a REST server, using the WebServer library of Vert.x. The received requests are forwarded through the EventBus to a specialized Handler Verticle, as the service is required. These Verticles can communicate with each other to perform extra functions (e.g. See if the device you are requesting statistics of a campaign is blocked) or simply send or request data to the communication module with the server database (using a MongoDB[13] DBMS as will be discussed later). Figure 2 shows a schematic view of this structure.

Another key point of this technology is its ability to be distributed, either by multiple processors, machines or even in different geographic locations, allowing having multiple servers around the level of a Wide Area Network (WAN) as a way of redundancy and more closeness to the points where there is greater flow of traffic, increasing overall throughput. Thus mobile devices can only communicate with a server/service, through a communication protocol. Since the client applications will work on mobile devices, it is expectable that the connection to the server can occasionally be completely absent or occasionally with a high degree of instability. It is also worth noting that this same connection can be chargeable. It is

[7]https://github.com/Adaptv/ribs2 (last visited in March 2015).

[8]http://vertx.io/ (last visited in March 2015).

[9]http://netty.io/ (last visited in March 2015).

[10]http://tomcat.apache.org/ (last visited in March 2015).

[11]https://github.com/eventmachine/eventmachine (last visited in March 2015).

[12]http://nodejs.org (last visited in March 2015).

[13]http://www.mongodb.org (last visited in March 2015).

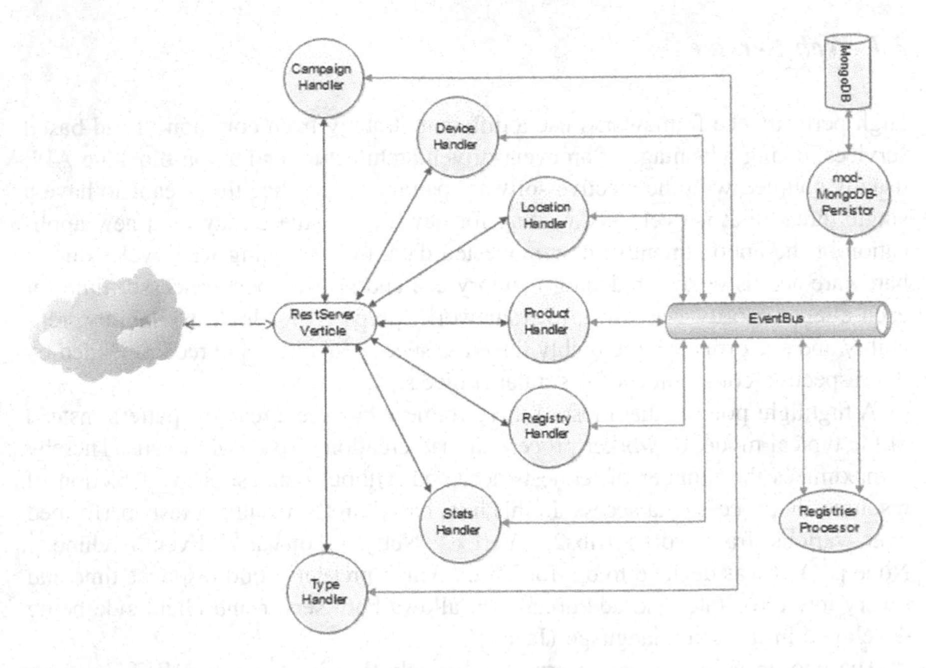

Fig. 2 Service structure

therefore important to choose simple protocols which do not require constant connection, having low overhead and a minimum of reliability. In addition it is also important that there are already tools and libraries available to support these communication protocols.

We therefore considered various client-server communication solutions, typically used in Web Services, since RPC (Remote Procedure Call) to more generic protocols (JSON-RPC, REST, CORBA, WebSockets, Eventsource, XML-RPC, SOAP). Through data obtained in tests and the knowledge gained from past projects, the REST protocol with a JSON structure was chosen in order to optimize as much as possible peering, facilitate debugging operations and increase the range of libraries for handling messages.

3.2 Crowd Sensing Database

In addition to framework's choice, the method to store data is also prepared to support large volumes of data in a distributed manner using specific DBMS for each end in the architecture. Importantly already mature relational databases and especially the new NoSQL databases that have been a hotly debated issue, both replication capacity, distributed processing and (case of NoSQL) supporting an unfixed data structure.

The Android framework itself already has support for SQLite, which has proven to have excellent performance in low concurrency and very low resource devices, these were the main reasons to automatically opt for this technology on the client.

Contrary to the client part, the service/server needs a database that supports plenty of concurrent accesses and that can effectively take advantage of available resources. Since there are various options, it was decided to refine the search with greater relevance to free databases, free use and good performance and reliability in parallel with the chosen framework. We decided also to investigate the use of NoSQL databases, which has lately been the subject of much discussion because of its proven more efficient ways to store information and better redundancy to prevent failure.

Centered on the concept of storing documents, Document model is designed precisely to save encapsulated data according to a standard, usually XML, JSON, BSON and PDF. Each document is assigned a unique ID throughout the database and sometimes some extra processing can be made, such as sorting and aggregations.

Within NoSQL databases (MongoDB and Cassandra[14]) it has been found advantageous to opt for MongoDB, mainly due to the flexibility of data, using JSON interface (such as the framework chosen), easy to use and ideal for applications prototyping process. On the other side, we excluded Cassandra due to its higher complexity and because it is a DBMS optimized for a much bigger data volume and infrastructure.

3.3 Mobile Application

The client-side of this platform (Fig. 3) allows a volunteer to register its device in a campaign already created on the server side. After that, the device is ready to submit new readings of donated goods by its barcode. In this way, no time or manual introduction is required, timestamp and geographical position are locally recorded. As soon Wi-Fi access is detected, the application transmits these records to the service populating the server database. As output, any volunteer who is participating in the same campaign can also see which amount of goods was donated so far. This is completely different from the old scenario where Caritas, only in the end of a week could account, and with no guarantees of accuracy, the success of a given campaign.

Specifically in the retail area, there is a known regulator GS1,[15] in order to define the type of barcode for each application and reserve blocks of barcodes in order to prevent collision of codes of different products. Although there are no technological restrictions, throughout this project the barcode EAN-13 format is the focus of study, since it is the most commonly used and easiest to test in Portugal. This

[14]http://cassandra.apache.org/ (last visited in March 2015).

[15]http://www.gs1.org/ (last visited in March 2015).

standard is public and present in all products sold in national retail outlets, identifying a type of product (not working as a serial number mechanism). Thus, as we are using a global format, which is public, it is clear that there will be a set of currently available tools that allow the reading and recognize these codes.

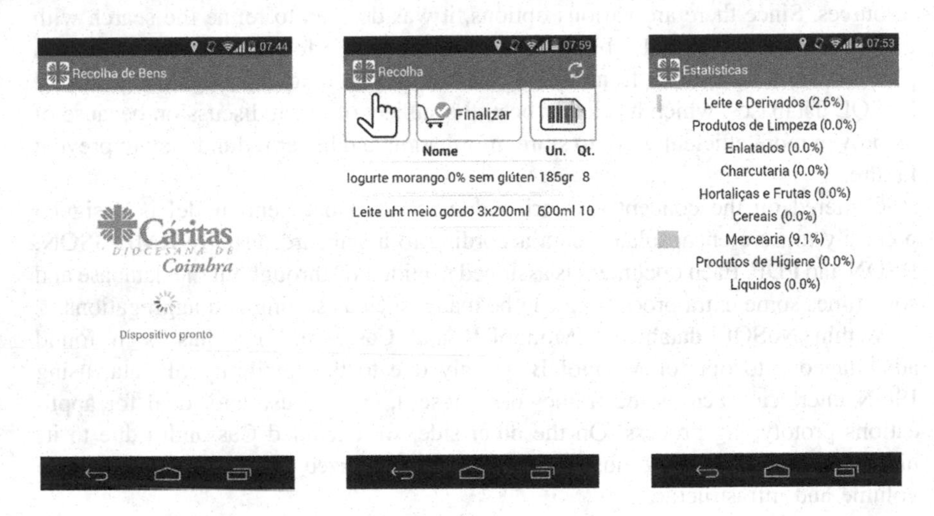

Fig. 3 Screenshots of the mobile application

After simple tests over two libraries (ZXing[16] and Scandit Barcode[17]), small prototypes were built to validate and evaluate the process and the ease of integration of those libraries in a real mobile application. Since the early research stage, the ZXing library proved to be the best accepted by most Android developers and since it is an entirely open source library and supports a wide range of 1D and 2D codes. The majority of Android applications require this library to read barcodes. The Scandit Barcode Scanner library is not open-source and has a much narrower range of supported codes (especially in the free version). However it provided better reading speed on less favorable luminosities and lower resolution cameras. Also there was the possibility of reading an EAN 13 code even with large inclinations of the camera and poorly focused image. With this, we managed to catalog a big amount of products, previously shorted by category and weight (reading one product by category + weight), adjust quantities and submit the data in just 8 min, a time much shorter than before.

[16]com.google.zxing.client.android (last visited in March 2015).

[17]com.scandit.demoapp (last visited in March 2015).

4 Conclusions and Future Work

In this paper we presented a Mobile Crowd Sensing platform in the context of solidarity campaigns. As active contribution from volunteers the platform receives in real time, recordings related to donations of goods that are in their majority identified by EAN barcodes. Thus, the mobile application is easy to operate and it allows seeing anytime the total amount received in a given campaign. The architecture proposed favors high availability and concurrency, managing different campaigns even geographically disperse. This project has just been set up in Caritas Diocesana of Coimbra, a Solidarity Institution and data are going to be collected through different campaigns. Nevertheless, this platform is of free use to other NGOs and is publicly available for open use.[18] We hope with this application being used by Institutions and final users we can collect intensive spatial and temporal data in order to recognize patterns in the giving process. We also plan to extend the platform and the mobile application to cover monetary donations since Caritas promote these national campaigns on the streets. This new mode of engagement offers opportunities to NGOs for reaching new donors under new circumstances as messages spread virally through friend networks.

References

1. H. Rheingold, *Smart Mobs: The Next Social Revolution* (Basic Books, New York, 2002)
2. F. Ye, R.K. Ganti, H. Lei, Mobile crowdsensing: current state and future challenge. IEEE Commun. Mag. **49**(11), 32–39 (2011)
3. D. Estrin, M. Hansen, A. Parker, N. Ramanathan, S. Reddy, M.B. Srivastava, J. Burke, Participatory sensing. in *Workshop on World- Sensor-Web, ACM SenSys* (2006)
4. T. Silvasti, G. Riches, *First World Hunger Revisited: Food Charity or the Right to Food?*, 2nd edn. (Palgrave Macmillan, Hampshire, 2014), pp. 72–86
5. A. Liang, L. Biderman, A. Ratti, C. Pereira, F. Oliveirinha, J. Gerber, A. Vaccari, A holistic framework for the study of urban traces and the profiling of urban processes and dynamics. in *12th International IEEE Conference on Intelligent Transportation Systems* (2009)
6. F. Rodrigues et al., Estimating disaggregated employment size from points-of-interest and census data: from mining the web to model implementation and visualization. in *International Journal on Advanced Intelligent Systems*, vol. 7 (2013)
7. W. Chen, T. Givens, Mobile donation in America. Mob. Media Commun. **1**(2), 196–212 (2013). doi:10.1177/2050157913476028
8. CITEK, Lyon, France (2014, September), http://www.yourinnovationday.eu/wp-content/uploads/2014/10/Presentation-CARITAS.pdf

[18]com.longinus.projrbandevida (last visited in March 2015).

3D Reconstruction of Bone Structures Based on Planar Radiography

Ana Coelho, João Pedro Ribeiro, Jaime Campos, Sara Silva and Victor Alves

Abstract The 3D reconstruction of bone structures has many advantages in orthopedic applications. 3D bone models could be used in computer assisted surgery systems or in the pre-operative planning of an orthopedic surgery. The visualization of these models will lead to higher surgery accuracy. Usually the 3D reconstruction is done with CT or MRI scans. However these modalities have some disadvantages like the high costs, high acquisition time and high radiation. So, the planar radiography emerges as a more advantageous modality, because it avoids exposure to high radiation, reduces the acquisition time and costs and also is the most usually acquired study in the pre-operative planning of an orthopedic surgery. The principal challenge in reconstructing bone models from planar radiography is that a lot of information is missing when only one or two orthogonal images are used. So it's hard to obtain a precise geometry of the bone structure with only this information. In this work, we present a solution for the problem of reconstructing bone structures from planar radiography. With this solution, it's possible to obtain a 3D model of the bone that is suitable for orthopedic surgery planning.

Keywords 3D reconstruction · Planar radiography · Orthopedic surgery planning

1 Introduction

The three-dimensional reconstruction of patient specific bone structures is necessary in a variety of medical applications, more specifically in orthopedic applications.

A. Coelho · V. Alves (✉)
Department of Informatics, University of Minho, Braga, Portugal
e-mail: valves@di.uminho.pt

J.P. Ribeiro · J. Campos · S. Silva
PeekMed, Braga, Portugal

A. Mohamed et al. (eds.), *Ambient Intelligence - Software and Applications*,
Advances in Intelligent Systems and Computing 376,
DOI 10.1007/978-3-319-19695-4_14

One example is pre- and intra-operative planning in Computer Assisted Surgery (CAS) systems that will lead to higher surgery accuracy [1]. An important part of these type of systems is to register 3D patient-specific bone models to in vivo data real-time, therefore during the pre-operative planning it is necessary to construct the 3D patient-specific bone model [2].

It's possible to point out some orthopedic surgeries that will benefit with pre-operative 3D visualization. One is the arthroplasty, both knee and hip, because with the 3D patient-specific bone model it's possible to customize the design of implant components. Another procedure is osteotomy, that can be performed to straighten a bone that has healed crookedly following a fracture or to correct a congenital deformity. Thus the 3D visualization of the patient-specific bone model can enhance the accuracy of this surgery. This type of models can also be used to personalize the treatment of pathologies such as scoliosis [4].

Usually 3D models are reconstructed from CT (Computed Tomography) or MRI (Magnetic Resonance Imaging) scans, however there are some disadvantages with these modalities. For example, CT induces high radiation doses to the patient, in fact it's even necessary to limit the number of scans to protect the patient [5]. Regarding MRI, this modality cannot be performed with metallic pieces, so when it's necessary to follow-up a patient with a metallic implant, it's not possible to do this with MRI scans [6]. Others disadvantages related with these modalities are their high costs, high acquisition time and also the acquisition of the scans in the supine position. This last disadvantage is an inconvenience for the clinical assessment of spinal deformities where it's necessary to acquire the scans in the natural standing position [4].

An alternative to these modalities is the planar radiography because this modality avoids exposure of the patient to high radiation, reduces the acquisition time and costs, and also allows the acquisition of the scan with the patient in the natural standing position [4].

In the planar radiography modality it's possible to have one 2D planar x-ray image or two 2D orthogonal images, called posteroanterior and lateral images. The x-ray image is just a projection of the patient's body [7], where all the organs and bones are superimposed. So the surgeon has to mentally visualize the anatomy of interest [3]. As such, the 3D reconstruction of x-ray images will be very helpful to the surgeon, because with this reconstruction he can actually visualize the bone from all directions and doesn't have to mentally visualize it. With the 3D model of the bone, the surgeon can examine the details that are missing in the x-ray images and plan the surgery in an easier and more efficient way.

A CT or MRI study has a number of slices, so a lot of information about the anatomy of the region of interest is available, however in x-ray studies with only one or two orthogonal images all this information is absent [8]. So it's hard to obtain a precise geometry of the bone structure with only this information.

1.1 State of the Art

The methods for 3D reconstruction of bone structures from 2D planar x-ray images can be divided in two types: the methods based only in the x-ray images and some a priori knowledge about topology and geometry of bone [9, 10, 13] and the methods based both in the x-ray images and in a template model of the bone structure that will be reconstructed [1–3, 5, 6, 8].

In the first type of reconstruction it's necessary to obtain calibration points that will be used as parameters for the reconstruction algorithms. These points can be obtained either by manual identification of anatomical landmarks or with a calibration object. This task when it's performed manually is particularly time consuming, tedious and susceptible to error [4]. When this is done with a calibration object there are also some problems because this object is used during the acquisition of the study and cannot be always available. Also it has a different configuration for each bone structure, being necessary to always change its arrangement.

After the identification of calibration points, it's necessary to obtain the 3D coordinates of those points. This can be done with the Direct Linear Transformation (DLT) algorithm for the stereo corresponding points, i.e. the points present in the two x-ray images [10]. For the non-stereo corresponding points, i.e. the points present in only one x-ray image, the Non Stereo Corresponding Points (NSCP) algorithm is used [9]. Once the 3D reconstruction of calibration points is done, the kriging algorithm is used to obtain the detailed model of the patient specific bone structure. This algorithm makes use of 3D coordinates of calibration points and some a priori knowledge about topology and geometry of the bone [9].

In the second type of reconstruction, the template model will complete the information that is missing in the x-ray images. The patient specific bone structure will result from the deformation of that model [8].

The template model of the bone structure is constructed with CT data of a population. It can be a statistical model constructed with Principal Component Analysis (PCA) [6] or a simplified personalized parametric model (SPPM) generated through descriptive parameters [11]. These descriptive parameters are defined according to the anatomical landmarks identified in the x-ray images [11]. However these methods involve the construction of a large database of models. A simpler method is to obtain the template model through the 3D reconstruction of a CT scan from a healthy subject of normal height and weight [12]. With this last method it's not necessary to construct a wide database of models, it's only necessary to have a model of each bone. However, it will increase the reconstruction time [13].

In this type of reconstruction the definition of calibration points is also needed. In most cases this task is performed manually but in some cases a semi-automatic method can be used [2, 14].

Once the template model is created it's possible to fit it to the patient specific bone structure. This is called the deformation step. This step can be performed using the descriptive parameters and the anatomical landmarks of the statistical model [11]. Other approaches are rigid transformations [18] and the Free Form Deformation (FFD) technique [19]. This last technique consists in embedding the model in a hull object that is deformable. Then, instead of deforming the model directly, the hull is deformed, and consequently the model inside is also deformed [19].

1.2 Proposed Solution

In this work, we present a solution for the problem of reconstructing bone structures from planar radiography. With this solution, it's possible to obtain a 3D model of the bone from planar radiography that can be used to perform surgery planning. The solution present in this work uses the second type of reconstruction, so a template model is used to generate the 3D model of the patient's specific bone structure. This type of reconstruction is more advantageous than the first type because the template model supplies the information that is missing in the x-ray images, so it's not necessary to define a large number of calibration points. Thus the calibration object is not necessary during the exam's acquisition and also it's not necessary to perform the 3D reconstruction of these points. In fact, in this type of reconstruction it's only necessary to define some calibration points, called anatomical landmarks, which will be used to register the template model with the x-ray images in order to obtain the deformation that is necessary to apply. The template model was created from the 3D reconstruction of CT scans from a healthy subject with normal height and weight. Therefore not only it's not necessary to construct a large database of bone models, but also only one model is needed for each bone structure.

2 3D Reconstruction

The 3D reconstruction of bone structures from planar radiography was organized into three main steps: generation of the generic model, 2D/2D registration and deformation of the generic model. In Fig. 1 are represented the various steps of the 3D reconstruction process.

These three steps will be explained in the next sections.

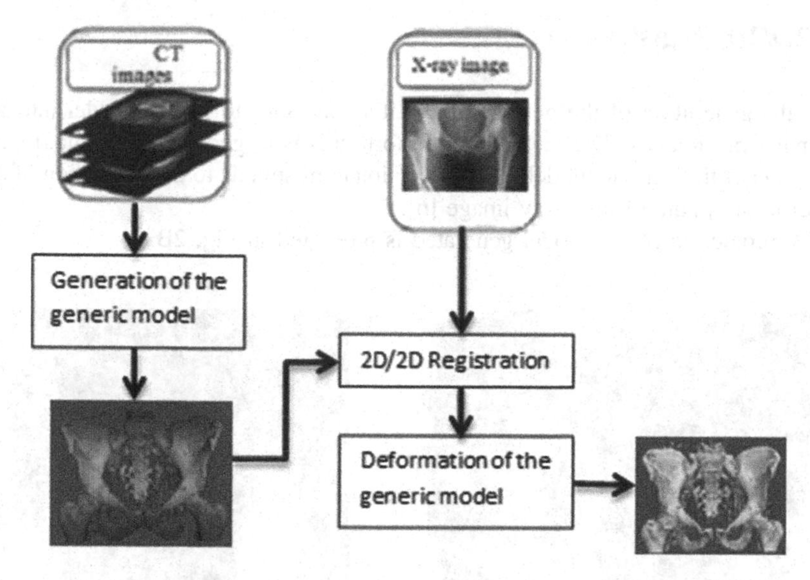

Fig. 1 Workflow for 3D reconstruction from planar radiography

2.1 Generation of the Generic Model

The first step to reconstruct the bone structure from planar radiography is to generate a generic model. This model was reconstructed from CT scans of healthy subject with normal height and weight. The generic model will then be deformed in order to fit the x-ray images. The deformed model will correspond to the 3D model of the patient's specific bone structure. The Marching Cubes algorithm was used to perform the 3D reconstruction of the generic model from CT scans [20].

This algorithm creates an isosurface from medical data with multiple 2D slices, such as the ones provided from CT or MRI scans. The isosurface created consists in a set of triangles extracted from each voxel of the study with an isovalue defined by the user [23]. The voxels with a value higher than the isovalue belong to the isosurface and the voxels with a value below don't belong to the isosurface.

The voxels of the study are represented as edges of a cube and by determining which edges of the cube are intersected by the isosurface, triangular patches can be created. This will divide the cube between regions within the isosurface and regions outside. The connection of the patches from all cubes on the isosurface boundary, allows obtaining a surface representation [20].

The generic model resulting of 3D reconstruction from CT scan with Marching Cubes algorithm is presented in Fig. 2A. In this work, we choose to perform the 3D reconstruction of the pelvic bone. All the figures of the models, projections and deformation field present in this work were visualized with the ParaView software [24].

3 2D/2D Registration

After the generation of the generic model it's necessary to fit this model into the original x-ray images. The first step to perform this is to generate the radiographic projection of the generic model. This projection corresponds to the projection of the model in the plane of the x-ray image [6].

The projection of the model generated is presented in Fig. 2B.

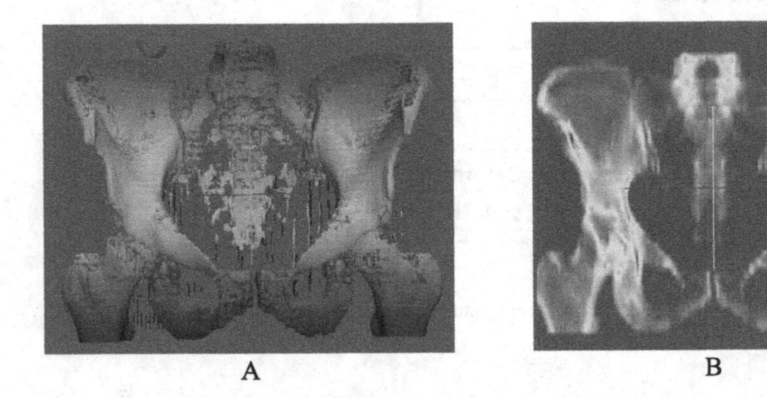

Fig. 2 Pelvic bone: A generic model; B projection of the generic model

Once the radiographic projection of the generic model is created it's possible to register this projection with the original x-ray image. This step called the 2D/2D registration is done to obtain the deformation that is necessary perform in order to fit the generic model with the x-ray image. The registration was performed with landmarks, so the first step consisted in the identification of anatomical landmarks in the projection and the original x-ray image. This identification was performed manually and according with the anatomical landmarks defined in the literature for the pelvic bone [21]. In Fig. 3A, it's possible to observe the x-ray image with the identified anatomical landmarks.

After the landmarks' identification, the coordinates identified in the two images were compared and a deformation field was created based on the differences of the coordinates [25]. This deformation field corresponds to the deformation that needs to be applied in the generic model in order to fit this with the original x-ray image.

In Fig. 3B, it's presented the deformation field superimposed on the projection of the model.

3.1 Deformation of the Generic Model

The last step to reconstruct the patient's specific bone structure from planar radiography is the deformation of the generic model. This was done with the deformation field created in the step of registration. The deformation field is an image of vectors, which means that every pixel contains a vector of displacement. So, the deformation is performed at a given point by adding the displacement at that point to the input point [26]. The final result is a deformed model that fits the original x-ray image, that is the 3D model of patient's specific bone structure.

The final deformed model is presented in Fig. 3C.

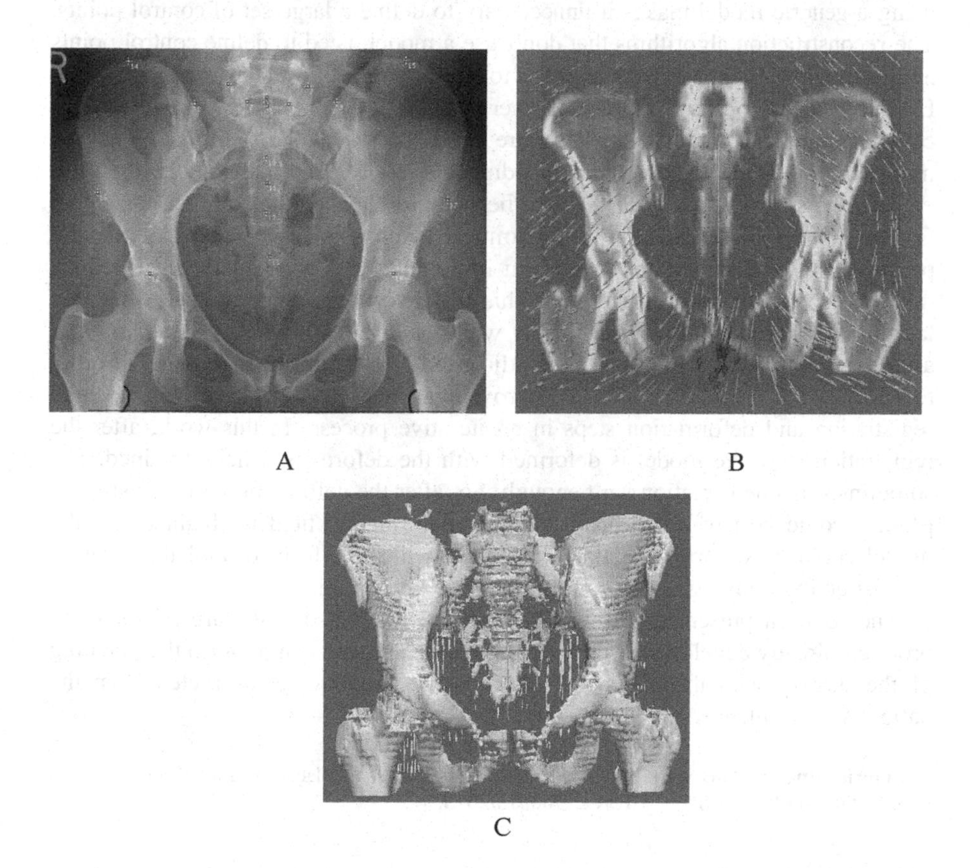

Fig. 3 Registration and deformation: A landmarks identified in the x-ray image; B deformation field; C final deformed model

4 Discussion

There are many advantages of reconstructing bone structures from planar radiography. The main advantages are related with the costs and acquisition time of this modality in comparison with the other modalities (CT and MRI) [4]. Another advantage is the fact that the planar radiography is the usually acquired study in the preparation of an orthopedic surgery. So, normally the surgeon only has the x-ray images of the patient to plan the surgery. Then, the 3D reconstruction from this modality is very helpful in the planning of orthopedic surgeries.

The solution presented in this work performs the 3D reconstruction of bone structures from planar radiography. With this solution it's not necessary to have a wide database of bone models, in fact it's only necessary to have one model of each bone. So, this solves the problem of having to construct large databases [8]. Also, using a generic model makes it unnecessary to define a large set of control points. The reconstruction algorithms that don't use a model, need to define control points and use their 2D and 3D coordinates to perform the 3D reconstruction [9, 10]. However, the algorithms that use a generic model need only a few control points, called anatomical landmarks, which are used to control the deformation of the model. In this work, the anatomical landmarks are identified manually by the user.

There is some future work that can be done to enhance the solution presented. For example, the identification of anatomical landmarks can be a semi-automatic process. This will reduce the time spent in this task and consequently will reduce the time spent for reconstruction. If this process were semi-automatic, first the landmarks were identified automatically with machine learning algorithms [14] and after this identification the user could adjust the position of the landmarks in order to enhance the precision. Another improvement that can be done is grouping the registration and deformation steps in an iterative process. In this work, after the registration step, the model is deformed with the deformation field obtained. But sometimes this deformation isn't enough. So, after the deformation, the registration process could be performed again, another deformation field is obtained and the model could be deformed again. The model will be deformed until the error of reconstruction converges or a limit number of times [22].

The solution presented in this work will be integrated with surgical planning products already developed [27]. In this way, the surgeon can perform the planning of the surgery with the 3D model of the bone structure reconstructed from the patient's x-ray images.

Acknowledgments This work has been supported by FCT - Fundação para a Ciência e Tecnologia within the Project Scope UID/CEC/00319/2013.

References

1. N. Baka, B.L. Kaptein, M. de Bruijne, T. van Walsum, J.E. Giphart, W.J. Niessen, B.P.F. Lelieveldt, 2D-3D shape reconstruction of the distal femur from stereo X-ray imaging using statistical shape models. Med. Image Anal. **15**(6), 840–850 (December 2011)
2. M. Mahfouz, A. Badawi, E.E.A. Fatah, M. Kuhn, B. Merkl, Reconstruction of 3D patient-specific bone models from biplanar x-ray images utilizing morphometric measurements, in *Proceedings of the 2006 International Conference on Image Processing, Computer Vision, and Pattern Recognition*, USA, 26–29 June 2006
3. P. Gamage, S.Q. Xie, P. Delmas, W.L. Xu, Diagnostic radiograph based 3D bone reconstruction framework: application to the femur. Comput. Med. Imaging Graph. 35(6), 427–437 (September 2011)

4. T. Cresson, R. Chav, D. Branchaud, L. Humbert, B. Godbout, B. Aubert, W. Skalli, J.A. De Guise, Coupling 2D/3D registration method and statistical model to perform 3D reconstruction from partial x-rays images data, in *Proceedings of the 31st Annual International Conference of the IEEE Engineering in Medicine and Biology Society: Engineering the Future of Biomedicine, EMBC 2009*, pp. 1008–1011 (2009)
5. W. Wei, G. Wang, H. Chen, 3D reconstruction of a femur shaft using a model and two 2D X-ray images, in *2009 4th International Conference on Computer Science and Education, IEEE*, pp. 720–722 (July 2009)
6. H.Boussaid, E.C. Paris, N. Paragios, 3D Model-based reconstruction of the proximal femur from low-dose biplanar x-ray images, pp. 1–10 (2011)
7. C. Koehler, T. Wischgoll, F. Golshani, Reconstructing the human ribcage in 3d with x-rays and geometric models. IEEE Multimedia **17**(3), 46–53 (2010)
8. S. Filippi, B. Motyl, C. Bandera, Analysis of existing methods for 3D modelling of femurs starting from two orthogonal images and development of a script for a commercial software package. Comput. Methods Programs Biomed. **89**(1) (January 2008)
9. R. Dumas, A. Le Bras, N. Champain, M. Savidan, D. Mitton, G. Kalifa, J.P. Steib, J.A. De Guise, W. Skalli, Validation of the relative 3D orientation of vertebrae reconstructed by bi-planar radiography. Med. Eng. Phys. **26**(5), 415–422 (June 2004)
10. B. Zhang, S. Sun, J. Sun, Z. Chi, C. Xi, 3D reconstruction method from biplanar radiography using DLT algorithm: application to the Femur, in *2010 First International Conference on Pervasive Computing, Signal Processing and Applications, IEEE* (2010)
11. S. Quijano, A. Serrurier, B. Aubert, S. Laporte, P. Thoreux, W. Skalli, Three-dimensional reconstruction of the lower limb from biplanar calibrated radiographs. Med. Eng. Phys. **35** (12), 1703–1712 (December 2013)
12. K. Koh, Y.H. Kim, K. Kim, W.M. Park, Reconstruction of patient-specific femurs using X-ray and sparse CT images. Comput. Biol. Med. **41**(7), 421–426 (July 2011)
13. E. Jolivet, B. Sandoz, S. Laporte, D. Mitton, W. Skalli, Fast 3D reconstruction of the rib cage from bi-planar radiographs. Med. Biol. Eng. Comput. **48**(8), 821–828 (August 2010)
14. R. Donner, B.H. Menze, H. Bischof, G. Langs, Global localization of 3D anatomical structures by pre-filtered Hough forests and discrete optimization. Med. Image Anal. **17**(8), 1304–1314 (December 2013)
15. P.E. Galibarov, P.J. Prendergast, A.B. Lennon, A method to reconstruct patient-specific proximal femur surface models from planar pre-operative radiographs. Med. Eng. Phys. **38** (10), 1180–1188 (December 2010)
16. Y. Chaibi, T. Cresson, B. Aubert, J. Hausselle, P. Neyret, O. Hauger, J.A. de Guise, W. Skalli, Fast 3D reconstruction of the lower limb using a parametric model and statistical inferences and clinical measurements calculation from biplanar X-rays. Comput. Methods Biomech. Biomed. Eng. **15**(5), 457–466 (January 2012)
17. M.K. Lee, S.H. Lee, A. Kim, I. Youn, T.S. Lee, N. Hur, K. Choi, The study of femoral 3d reconstruction process based on anatomical parameters using a numerical method. J. Biomech. Sci. Eng. **3**(3), 443–451 (2008)
18. X. Dong, G. Zheng, Automatic extraction of proximal femur contours from calibrated X-ray images using 3D statistical models: an in vitro study. Int. J. Med. Rob. Comput. Assist. Surg. **5**, 213–222 (April 2009)
19. S. Filippi, B. Motyl, C. Bandera, Comparing parametric solid modelling/reconfiguration, global shape modelling and free-form deformation for the generation of 3D digital models of femurs from X-ray images. Comput. Methods Biomech. Biomed. **12**(1), 101–108 (2009)
20. W.E. Lorensen, H.E. Cline, Marching cubes: a high resolution 3d surface construction algorithm. SIGGRAPH Comput. Graph. **21**(4), 163–169 (August 1987)
21. S. Delorme, Y. Petit, J.A. de Guise, H. Labelle, C.E. Aubin, J. Dansereau, Assessment of the 3-d reconstruction and high-resolution geometrical modeling of the human skeletal trunk from 2-D x-ray images. IEEE Trans. Biomed. Eng. **50**(8) (2003)
22. T. Cresson, B. Godbout, D. Branchaud, R. Chav, P. Gravel, J.A.D. Guise, Surface reconstruction from planar x-ray images using moving least squares, in *30th Annual*

International IEEE EMBS Conference Vancouver, British Columbia, Canada, 20–24 August 2008, pp. 3967–3970 (2008)
23. G.M. Nielson, B. Hamann, The asymptotic decider: resolving the ambiguity in marching cubes, in *Proceedings of the 2nd conference on Visualization'91*. IEEE Computer Society Press, pp. 83–91 (1991)
24. ParaView 4.3.1. http://www.paraview.org/. Accessed 4 Dec 2014
25. ITK LandmarkDisplacementFieldSource. http://www.itk.org/Doxygen45/html/classitk_1_1LandmarkDisplacementFieldSource.html. Accessed 15 Dec 2014
26. ITK DisplacementFieldTransform. http://www.itk.org/Doxygen45/html/classitk_1_1DisplacementFieldTransform.html. Accessed 15 Dec 2014
27. J. Ribeiro, V. Alves, S. Silva, J. Campos, A 3D computed tomography based tool for orthopedic surgery planning, in *Developments in Medical Image Processing and Computational Vision, 2014* (2014)

Context-Aware Well-Being Assessment in Intelligent Environments

Fábio Silva, Celestino Gonçalves and Cesar Analide

Abstract The implementation of concepts such as smart cities, ambient intelligence and internet of things enables the construction of complex systems that may follow users across environments through many devices. One potential application is the assessment and assurance of well-being of users within different environment with different configurations. This is a complex task that requires the capture of the state and context of both users and environments through sensors dispersed across environments and users. It's the opportunities created by the emergence of technology that provide enough information to intelligent autonomous systems. Adapting expectations of a well-being assessment system to task and context is possible using the new techniques imported from different fields such as sensor networks, sensor fusion and machine learning. This article encompasses the design and implementation of a platform to evaluate well-being according to each context and translate it to sustainable indicators.

Keywords Sensors networks · Ambient intelligence · Sustainable indicators · Well-Being

1 Introduction

The internet of things is a new paradigm, in which every device is digitally connected, regardless of their function and communicates with other devices and other people over communication protocols. It applies both to fixed devices and personal devices that accompany people [1]. More examples can be enumerated by devices that are being incorporated inside the actual body, such as identification chips,

F. Silva (✉) · C. Gonçalves · C. Analide
Department of Informatics, University of Minho, Braga, Portugal
e-mail: fabiosilva@di.uminho.pt

C. Analide
e-mail: analide@di.uminho.pt

A. Mohamed et al. (eds.), *Ambient Intelligence - Software and Applications*,
Advances in Intelligent Systems and Computing 376,
DOI 10.1007/978-3-319-19695-4_15

smart tattoos and alike [2]. Smart city is a term applied to digital research using computational methods and systems that results in better, easier and faster management of services and goods inside inhabited areas. In this setup, the internet of things acts as a base service which enables smart cities applications to collect information directly from the environment and people and the integrated fusion of data and information. This benefits the planning of actions to improve the status quo. Among other concerns, health, comfort and well-being are topics being addressed in smart cities research [3].

If the technologies described under the concepts of smart cites and internet of things are perceived as social services, then it is possible to gain access to a new set of valuable information on both the environment and users. These trends, despite having ethical challenges of their own, present a number of opportunities for society. Sensorization, monitoring and sharing of information are terms intimately connected to the new intelligent systems being created. Even more, applied research related to health organizations and also the well-being of populations or individuals is present in recent studies. Connected environments and the act of monitoring comfort parameters are not under active research but also regulated by governments. For examples, air quality is an area that is actually regulated by governments which define acceptable parameters. Concurrently, research conducted in the field of smart environments studies the impact of air composition in health, concentration tasks and psychological comfort. As expected, research directions are more specialized than government regulations and are being pushed forward by the quality of sensors and sensor networks which portrait better and better images of air composition across time and space inside environments. There are other examples with equal strategies that aim to better assess and diagnose unoptimised or harmful situations with focus on well-being and general health [4].

The quest for physical well-being is addressed under smart cities by the use of indicators from personal devices such as smartwatches, sport monitors or smart bands which among other things can monitor some health parameters such as heart rate, blood oxygen levels and sport activities. Research in the body response to environment parameters is also potentiated by these personal devices that act as personal sensors [5]. Their use to calculate more intrusive parameter such heat balance index for each individual helps perceive whether it is cold or warm in different situations. The environment condition alone is correlated to well-being and comfort status [6]. More importantly, there are decisions that can be made using such information that may affect both physical and psychological health. Using fixed sensors over specialized areas are of interest to assess environments and their impact on heath [7]. For instance, projects that monitor city or indoor air quality provide rich information about potential health risks that may impair both physical and psychological well-being. What is more, the availability of digital services allows a faster identification and actuation upon these threats. The difficulty gathering and constantly monitor such parameters leads to situations where interventions are planned later than they should and the problems worse. Social sustainable indicators developed to assess development of countries and populations can also be automated using sensory networks with devices directly connected with each other that can provide information that would normally require expansive survey, and field tests. In the health care community there are social indicators as well.

2 Related Work

According to the Oxford Dictionary of English, well-being is the state of being comfortable, healthy, or happy. In order to assess it, the three subjects should be covered by the process conducting such evaluation.

In a professional environment, according the World Health Organization a healthy job is likely to be one where the pressures on employees are appropriate in relation to their abilities and resources, to the amount of control they have over their work, and to the support they receive from people who matter to them. As health is not merely the absence of disease or infirmity but a positive state of complete physical, mental and social well-being, a healthy working environment is one in which there is not only an absence of harmful conditions.

Environment and physical working conditions are important organizational risk factors [8]. Previous approaches to this work focused on the physical attributes predicament such as heart rate, and comfort due to environment conditions. The assumption that comfort equilibrium calculation was too complex to perform on a recurrent analysis [9] are mostly proved wrong in an intelligent environment equipped with smart sensors. Although, some of the algorithms might still be considered complex, most mathematical can be used fairly easy by computational systems. From this perspective it is possible to perform scoring systems to rate the degree of comfort a user experiences. Traditional studies use ergonomic research to estimate ideal parameters that are adequate to the space, the people and the task performed. Considering these variables, in this study, we can analyse traditional attributes such as temperature, humidity, luminosity, relevant attributes like CO, CO_2 and airborne dust. However, it is possible to increase the precision of these models, not only estimating individual assessments but also correlated assessments through dedicated indicators with sensor and data fusion techniques. As an example there are a number of indexes that can be created, namely, perfected temperature, percentage of people in discomfort and stress related with heat perception [10] and adequate luminosity for each task. These sensors monitor the environment and its attributes looking for suboptimal attributes, however relate these attributes with user discomfort or user stress is still a relative study. There is another layer, physical human sensors which, in this case, monitor the heart rate for each individual inside smart environments. Whenever suboptimal values are detected, it will be possible to relate the physical state of each person to the environment attributes. Psychological and physiological adverse ambient conditions can produce significantly changes in a person. However authors orient this topic to a set of variables. In this particular approach sound, temperature and luminosity are studied as external factors that affect well-being and mood states. Previous authors have debated the influence of such variables in the impact of mood change in people [11]. Related studies between stress and well-being and stress recognition can also be found. In [12] we can see a wearable system for ambulatory stress monitoring recording a number of physiological variables known to be influenced by stress. In [13] author gathers values from a skin temperature sensor, a heart rate sensor and a skin conductance sensor. The signals from the sensors are input into a microcontroller where all the processing takes place and carried out though ZigBee technology. Data are stored in a computer it is stored for data analysis and feature extraction for emotion recognition.

Comfort is subjective directly related to a person's personality, beliefs and habits. There are however ergonomic studies that provide the necessary background to create an environment which satisfies the most common needs to keep both the people and environments healthy. Although not being an extensive review of possible attributes it does indicate a minimum set of possible attributes to start comfort and well-being assessments [14].

3 Context-Aware Well-Being System Architecture

A combination of personal and environmental attributes provides a better representation of the information required to assess well-being and hypotheses about user condition, its relation to environment attributes and the activities being performed. Such information though not being medically considered as hard prove to diagnosis may be used as soft information about condition and habits of living. This system depends on the number of sensing devices and attributes considered to make decisions more accurate and expressive. It is also important to adapt the notions of comfort and safety according to the context of tasks and activities being performed. Taking knowledge from a knowledge base about each environment, for instance relations between sensor and event records, it is actually possible to estimate the state of well-being, both currently and in the future.

3.1 Environment Evaluation

A home environment is a particular setup, with a very personal context. Depending on job state, a person may spend most of the time in this environment or only after work hours. Taking into consideration a typical residence with working people, it is possible to assess comfort condition and monitor and predict values of comfort.

From the installation of different sources of sensorization through the environment it is possible to measure the impact of each individual attribute in the environment. As a personal environment those should be the best suited to assess what each user deems as comfortable. However as users can have different meanings for comfort, there might exist some exceptions to this assertion. An environment, though not always equal it is generally categorized by a population of individuals sharing the same environment. Each environment has different requirements according to the specific tasks its users perform. An experience to demonstrate the findings of the system proposed was designed with a sensors network in place to monitor environment variables. During this experiment a presence control was instituted to add user presence as an attribute. The description of the summary of the data gathered is available in Table 1.

Table 1 Summarized data retrieved from the environment sensor network

	Mean	Standard Deviation	Max	Min
Temperature	19.98 °C	1.68 °C	24.88 °C	17.99 °C
Luminosity	122.53 lux	102.49 lux	632.00 lux	0.00 lux
Humidity	45.69 %	7.39 %	63.19 %	38.86 %
Number of people	2.27	1.70	10.00	0.00

The analysis of well-being is inherently different from home and professional configurations due to the fact that the context alters comfort values. Though some of the alterations remain within the acceptable range they can be used to introduce differences in the well-being analysis.

3.2 Personal Assessment

Well-being through environment alone is an incomplete study as it disregards the effects of user interaction and user behaviour. Under normal circumstances, well-being should generate values inside the satisfactory range for attributes being measured. Considering one individual alone, the existence of normal range values for environment and abnormal for the personal sensing might lead to the suspicion of something not right with the individual thus impairing well-being.

On Table 2, a set of indicators gathered from a personal sensorization hardware reveals the normal range of attribute date for a given individual. His historic data reveal how comfortable he is by assessing most common values after some period of time. As the indicator values go up or down the mean and standard deviation being considered it detects abnormal behavioural pattern and uses majority votes to decide whether it is really an uncomfortable behaviour or not.

A practical validation test can be made using records from environment and testing them against such data. It was perceived that environment conditions only directly affect personal attributes the most when they differ significantly under normal circumstances that is avoiding considerations about individual's state of mind. Environment variables ranging inside the comfort zone are less likely to produce chances in the normal values from personal sensorization.

Table 2 Summarized data retrieved from the user sensor network

	Mean	Standard deviation	Max	Min
Steps	4177.80	7537.96	5080.00	3245.00
Blood pressure	78.42 bpm	19.11 bpm	125.00 bpm	58.00 bpm
Oximetry	97.45 %	1.02 %	100.00 %	95.00 %
PMV	−0.79	0.86	1.20	−1.72

3.3 Well-Being Assessment

Well-being assessment is employed in a two-phase strategy. First, the analysis of critical conditions through environment sensory data and secondly, the analysis of comfort. At the initial phase, a thorough analysis about each individual attribute is made to make sure that each of the sensorized values are non-risk values towards human. This evaluation is made both to environment and personal attributes as shown in Table 3. It is important to deal with information quality and validity. If the evaluation fails at this stage the uncomfortable setup is immediately generated. Even if there is only one of the attributes outside what is considered safe range, the resulting classification is deemed non satisfactory and a 0 % well-being is issued as life might be endangered.

Following on to the second phase, with each user profile and environment profile, a match between activities and each attribute available is analysed. In the sample test, the temperature and luminosity of environments where chosen. Different comfort rules are created according to the dominant activities in each environment and the ideal configuration is assessed through ranges of values from medical, thermal and ergonomic studies present in configuration databases in the system. From this point, a weighted percentage is loaded from the database to allow different comfort attributes to have more or less impact in the well-being value. The final value of well-being is the weighted majority of the satisfaction comfort for each attribute. A percentage, less than 100 % denotes that at least some attribute being measured is not within comfort values for a given time. Finally, a daily assessment considers the average of well-being satisfaction between measures.

Table 3 Comfort Reference Values

Comfort Attribute	Lower Limit	Upper Limit
Humidity	20 %	60 %
Temperature	18 °C	25 °C
Luminosity	50 lux	107.527 lux
Temperature perception (PMV)	−1.5	1.5
Heart Rate (Woman)	60	78
Oximetry	90 %	100 %

Table 4 presents the relative percentages of time the environment was considered good for well-being according to the algorithm described with the analysis of the attributes and their matching to each other. It is shown that environment configurations are generally portrayed as good in well-being standing but the personal attributes are lacking more often. In the case of this experiment that is due to non-activity, and sedentary lifestyle. Its implications are that, although the environments appear well configured, the activities of users are not being considered as good for well-being. Although with this setup, if the conditions verified are not directly correlated between both environment and personal attributes, the weighted average among all attributes decides the category of well-being.

Table 4 Well-being assessment experiment analysis

	Environment well-being	Personal well-being	Well-being assessment
Environment	97 %	78 %	83 %

Historical data about environment and both environment and people can be addressed making up profiles of aggregated information through the use of dashboards. These dashboards should contain valuable information for healthcare institutions and its professionals to help with patient consult for instance. More than the experience a verbal consult with a patient can produce, a virtual conversation is something that includes virtual data according to sensing system available. Though the system may be assessed with simulation data, it is in the real world data applied to each scenario that it usefulness can be extensively validated. For healthcare institution it is possible to navigate, through time windows and select only values of selected attributes and obtain both a measurement dashboard as well as a well-being classification. The input in the system are dependent in the number of attributes measured by the sensor network in place and the capability of its devices. The further specialized the devices and information is the further the quality of information and the validity of these analysis.

4 Conclusion

The development of our well-being as human beings is something that should not be overlooked. Like other important subjects to our daily life, continuous development and improvement is desirable and an objective. Systems such as these are starting to appear through the research community and as exploratory projects within large enterprises. While not identifying themselves as health organizations, their goal is to promote the investigation of well-being through the population whilst increasing the information available. Consequently, this leads to building large databases about user behaviour and environment conditions. This also starts the study of communities and the impact of behaviours and environment on health and comfort through large sets of population strengthening existent knowledge with large validation test, but most importantly creating the opportunity to generate new and improved knowledge by analysing these records. Seemingly close behaviour in different conditions can generate different results, it can be comparable to the butterfly effect, where a small change can have great implication on the whole system towards a positive or negative result. The categorization of these conditions may improve not only well-being but health as well. The promises and the large research body around these themes indicates that these technologies and these research considerations are valid. They are even today the object of pursuit by society as a mean to improve living conditions.

A future approach to this study should include distribution analysis between the historical data on a given time window and the sensorized values. This envision may also help annotate different states of comfort through building and the people inside to better map how the environmental attributes of such buildings affect the well-being of people. As tried in the experimental framework to detect stressful moods (Silva et al. 2013), where a testbed was used to perceive different room states during the day and the classification of probable well-being.

Acknowledgments This work was developed in the context of the project CAMCoF - Context-aware Multimodal Communication Framework funded by ERDF - European Regional Development Fund through the COMPETE Programme (operational programme for competitiveness) and by National Funds through FCT - Fundação para a Ciência e a Tecnologia within project FCOMP-01-0124-FEDER-028980 and PEst-OE/EEI/UI0752/2014. Additionally, it is also supported by a doctoral grant, SFRH/BD/78713/2011, by FCT in the financial program POPH/FSE in Portugal.

References

1. L. Atzori, A. Iera, G. Morabito, The internet of things: a survey. Comput. Netw. **54**(15), 2787–2805 (2010)
2. R. Steele, A. Clarke, The internet of things and next-generation public health information systems. Commun. Netw. **05**(03), 4–9 (2013)
3. A. Solanas, C. Patsakis, M. Conti, I. Vlachos, V. Ramos, F. Falcone, O. Postolache, P. Perez-martinez, R. Pietro, D. Perrea, A. Martinez-Balleste, Smart health: a context-aware health paradigm within smart cities. IEEE Commun. Mag. **52**(8), 74–81 (2014)
4. G. Piro, I. Cianci, L. A. Grieco, G. Boggia, P. Camarda, Information centric services in Smart Cities. J. Syst. Softw. **88**, 169–188 (2014)
5. M. Chan, D. Estève, J.-Y. Fourniols, C. Escriba, E. Campo, Smart wearable systems: current status and future challenges. Artif. Intell. Med. **56**(3), 137–156 (2012)
6. R. Rana, B. Kusy, R. Jurdak, J. Wall, W. Hu, Feasibility analysis of using humidex as an indoor thermal comfort predictor. Energy Build. **64**, 17–25 (2013)
7. L. Atallah, B. Lo, G.-Z. Yang, Can pervasive sensing address current challenges in global healthcare? J. Epidemiol. Glob. Health **2**(1), 1–13 (2012)
8. C. Biron, H. Ivers, J.-P. Brun, C.L. Cooper, Risk assessment of occupational stress: extensions of the clarke and Cooper approach. Health. Risk Soc. **8**(4), 417–429 (2006)
9. P.O. Fanger, *Thermal Comfort: Analysis and Applications in Environmental Engineering*. (Danish Technical Press, Copenhagen 1970)

10. K. Parsons, *Human Thermal Environments: The Effects of Hot, Moderate, and Cold Environments on Human Health, Comfort and Performance.* (Taylor & Francis, London, 2010)
11. K.C. Parsons, Environmental ergonomics: a review of principles, methods and models. Appl. Ergon. **31**(6), 581–594 (2000)
12. J. Choi, B. Ahmed, R. Gutierrez-Osuna, Ambulatory stress monitoring with minimally-invasive wearable sensors. Comput. Sci. Eng. Texas A&M, (2000)
13. M. Tauseef, Human Emotion Recognition Using Smart Sensors, Massey University, 2012
14. F. Silva, T. Olivares, F. Royo, M.A. Vergara, C. Analide, Experimental Study of the Stress Level at the Workplace Using an Smart Testbed of Wireless Sensor Networks and Ambient Intelligence Techniques, in *Natural and Artificial Computation in Engineering and Medical Applications SE - 21*, ed. by J. Ferrández Vicente, J. Álvarez Sánchez, F. de la Paz López, F. J. Toledo Moreo, vol. 7931 (Springer, Berlin Heidelberg, 2013), pp. 200–209

An Overview of the Quality of Service in Bluetooth Communications in Healthcare

Ana Pereira, Eliana Pereira, Eva Silva, Tiago Guimarães, Filipe Portela, Manuel Filipe Santos, António Abelha and José Machado

Abstract Currently, the general public requires devices getting faster and great performance, that is, devices ensuring a better quality of service. One way to achieve these goals is through the use of devices supported by the mobile computing with tools to help the search for information. Bluetooth technology is an open standard for wireless communication allowing the transmission of data and information between electronic devices within walking distance, with minimum resource expenditures, safe and rapid transition of data. So, the Bluetooth technology was initially designed to support simple network devices and personal devices such as mobile phones, PDAs and computers, but quickly it were discovered other applications in several areas. In this article, it will be performed a

A. Pereira · E. Pereira · E. Silva · T. Guimarães
University of Minho, Minho, Portugal
e-mail: a58539@alunos.uminho.pt

E. Pereira
e-mail: a58508@alunos.uminho.pt

E. Silva
e-mail: a60196@alunos.uminho.pt

T. Guimarães
e-mail: a61774@alunos.uminho.pt

F. Portela (✉) · M.F. Santos · A. Abelha · J. Machado
Algoritmi Research Centre, University of Minho, Minho, Portugal
e-mail: cfp@dsi.uminho.pt

M.F. Santos
e-mail: mfs@dsi.uminho.pt

A. Abelha
e-mail: abelha@di.uminho.pt

J. Machado
e-mail: jmac@di.uminho.pt

A. Mohamed et al. (eds.), *Ambient Intelligence - Software and Applications*,
Advances in Intelligent Systems and Computing 376,
DOI 10.1007/978-3-319-19695-4_16

literature review on the topic, with the goal to understand how the Bluetooth technology can benefit increases in the Quality of Service and the presentation of some actual and potential biomedical applications.

Keywords Bluetooth · Quality of service · Piconets · Master · Slave · Heal · Biomedical applications · Ubiquitous devices

1 Introduction

The Bluetooth technology is considered a low power technique, low cost and secure. It consists of a wireless communication technology, specially designed to replace the wires that interconnect and ubiquitous electronic devices that are relatively next, such as mobile phones or dial-up network. In addition, also fits printers, keyboards, headsets, and etc. To date, it has been considered a Personal Area Network (PAN) for ad hoc and network infrastructure [1]. More now than ever, with the release of the Core Specification 4.x, including the single-mode Bluetooth low-energy (BLE) technology or Bluetooth Smart) and dual-mode (both classic Bluetooth and BLE combined) or Bluetooth Smart ready, Bluetooth is a leading candidate to connect the Internet of Things.

The Bluetooth specification defines a radio frequency interface and a set of communication protocols for the discovery of devices, data exchange and bug fixes. By the other hand, the connection speed, communication range and the level of Bluetooth transmission also are chosen. So, Bluetooth is considered a low cost and consumption technology and thus has quickly become one of the most interesting technologies for communication within small distances. However, problems related with the Quality of Service (QoS) should be treated in order to ensure that these requirements are met by the system components of communication during data transmission. In networks without wire, the complexity increases to meet these requirements due to the high dynamics of the environment and the diversity in the quality of the links, caused mainly by the interference by high error rates and the mobility of users.

Commonly, in health units, mobile devices are relatively close to each other and usually around the patient. In this sense the main propose of this article is to understand how the Bluetooth technology can benefit the increase of QoS in a real context in a health institution (Centro Hospitalar do Porto). In this paper, a review of literature on the subject will be held in order to understand how Bluetooth technology can increased QoS in health institutions, patient recovery and their wellbeing. Thus, this study is the start point of a project that is being developed in the CHP institution, whose purpose is to introduce wireless communication between various devices attached to patient recovery, thereby improving QoS.

2 Background

2.1 Centro Hospitalar of Porto and AIDA

In Centro Hospitalar of Porto (CHP) sometimes the existence of many devices connected to the patient and the existence of wires becomes uncomfortable for the patient and for their recovering. In this sense the goal of this study is to understand how the Bluetooth technology can be a benefit by increasing the QoS at CHP and the overall quality patient life. In the case of CHP is implemented the (Agency for Integration, Archive and Diffusion of Medical Information). This is a platform based on multi-agent system. The main goal is to overcome the difficulties presented by the uniformity of clinical systems and by the complexity of medical and administrative data provided by different hospital data sources [2–4]. In addition and in this context, the AIDA emerges as one example of the technologies that may benefit from the full capabilities of Bluetooth and its subsequent increase of QoS [5].

2.2 The Bluetooth Technology

In the Bluetooth Core Specification version 4.0, the Bluetooth Special Interest Group (SIG), included, not only the well-known and widely used Classic Bluetooth (BT) but also the Bluetooth Smart or Bluetooth Low-Energy (BLE) introducing a brand new way of allowing the existence of a new generation of devices that can run for months without a recharge [15]. However, this has important limitations as well as benefits. It is quite different from Classic Bluetooth technology—so different that must carefully be considered which technology best fits the application needs [16]. Bluetooth is a set of specifications for common short-range wireless applications. These specifications include core components, and the application profiles that use them. These specifications are rigorously validated by the Bluetooth SIG before they are adopted and published and that these specifications are tested and qualified to ensure interoperability [15].

With the introduction of Bluetooth low energy technology, there has been considerable interest in its possibilities regarding many aspects present in the daily aspects of the users' life. It was born to allow the existence of small and simple devices that can be peripherals for smartphones and other devices, these are called Bluetooth smart devices. The main devices in which they are connected are called Bluetooth Smart Ready devices. Bluetooth Smart devices include only a single-mode low energy Bluetooth v4.0 radio. This Bluetooth Smart Ready devices are able to connect to both BT and BLE implementing a Bluetooth v4.0 dual mode radio, where Bluetooth low energy functionality is integrated into an existing Classic Bluetooth controller.

BLE features a very efficient discovery and connection set-up, short data packages, and asymmetric design for battery operated sensor types of applications.

As with Classic Bluetooth technology, Bluetooth low energy technology is based on a master connected to a number of slaves. However, in Bluetooth low energy technology the number of slaves can be as big as the implementation and available memory allows it to be. The new advertising functionality makes it possible for a slave to announce that it has something to transmit to other devices that are scanning. Advertising messages can also include an event or a measurement value.

There are also differences in software structure (Fig. 1). In Bluetooth low energy technology all parameters have a state that is accessed using the attribute protocol. Attributes are represented as characteristics that describe signal value, presentation format, client configuration, etc. The definitions of these attributes and characteristics along with their use make it possible to build several basic services and profiles like proximity, battery, automation I/O, building automation, lighting, fitness, and medical devices. All these nuances are needed to assure the implementation is compatible between devices from different manufacturers. While Bluetooth 2.0 EDR or Bluetooth 3.0 HS introduced a higher data rate functionality, Bluetooth 4.0 targets effective communication with devices that do not need streaming data or high data throughput. Low-energy requirements stipulate noticeable differences in Bluetooth and Bluetooth LE protocols, but at the same time take great advantage in reuse of the RF part, leaving most of the changes to the protocol stack [15, 18].

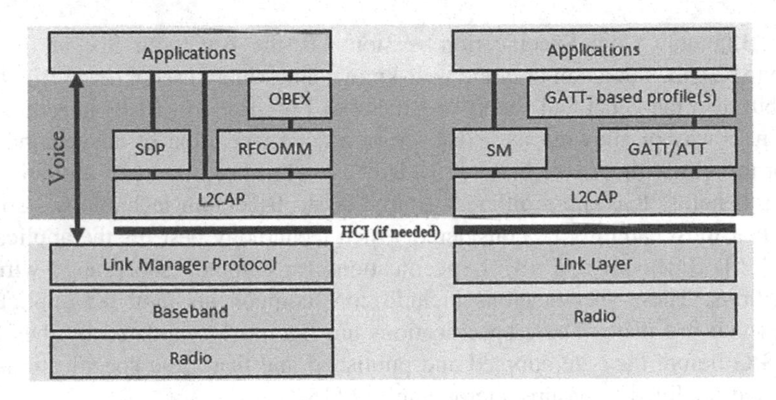

Fig. 1 A simplified version of the Classic Bluetooth and the BLE protocols

Although having some common layers, some major changes can be observed by the extinction and the rise of some parts and by the adjustment of others. Even though both have a radio layer, which promotes modulation and demodulation the physical layer signal, they are not the same. The frequency range is the same (in the 2.4 GHz band), they both use frequency hopping, and the modulation scheme is also equal (Gaussian Frequency Shift Keying (GFSK) modulation), however the modulation index is different (0.5 in BLE and 0.35 in BT). This change lowers power consumption and also improves the range of BLE versus Classic Bluetooth [6, 18]. The Classic Bluetooth includes a Baseband layer that transmits the packets directly and supports two types of links: Synchronous Oriented Connection (SCO) and Asynchronous not Oriented Connection (ACL). The SCO link has characteristics that can be found in circuit switching, while the connection ACL is more akin to packet switching. Includes additionally the LMP (Link Manager Protocol) that provides the basic functions for configuration and management ACL links. In Bluetooth LE is present the link layer (LL) controller which is responsible for low level communication over a PHY interface. It manages the sequence and timing of transmitted and received frames, and using a link layer protocol, communicates with other nodes regarding connection parameters and data flow control. It also handles frames received and transmitted while the device is in advertising or scanner modes. It also provides gate keeping functionality to limit exposure and data exchange with other devices. If filtering is configured, the LL controller maintains a "white list" of allowed devices and will ignore all requests for data exchange or advertising information from others. It not only helps with a security aspect but also reduces power consumption. Uses HCI to communicate with upper layers of the stack if they are not collocated [6, 18]. In both implementations can be found the HCI (host controller interface) layer that acts as a boundary between the lower layers of the Bluetooth protocol stack and the upper layers.

The Bluetooth specification defines a standard HCI to support Bluetooth systems that are implemented across two separate processors. For example, a Bluetooth system on a computer might use a Bluetooth module's processor to implement the lower layers of the stack but then use its own processor to implement the upper layers. In this scheme, the lower portion is known as the Bluetooth module and the upper portion as the Bluetooth host. However it is not required to divide the Bluetooth stack in this way. Some devices can combine the module and host portions of the stack on one processor, being this intended for small and self-contained devices. In such devices, the HCI may not be implemented at all unless device testing is required. The logical link control and adaptation layer protocol (L2CAP) component provides data services to upper layer protocols. It is responsible for protocol multiplexing data between different higher layer protocols and Segmentation and reassembly of packets, and de-multiplexing and reassembly operation on the other end [6, 18]. The upper layers are another aspect that marked a difference among the BT and the BLE. They consist of protocols allowing to discover services provided by other Bluetooth devices in the surrounding area of the device, and making sure it is done safely and efficiently. The Security manager in BLE and RFCOMM and OBEX in BT are examples of these layers. Bluetooth 4.0 introduces also a new communication method, called the attribute protocol (ATT) which is optimized for small packet sizes used in Bluetooth low energy. The ATT allows an attribute server to expose a set of attributes and their associated values to an attribute client. These attributes can be discovered, read, and written by peer devices.

The generic attribute profile (GATT) describes a service framework using the attribute protocol for discovering services, and for reading and writing characteristic values on a peer device. It interfaces with the application through the application's profiles. The application profile itself defines the collection of attributes and any permission needed for these attributes to be used in the communication [18].

3 Quality of Service

3.1 Concept

The concept of QoS has evolved over time. Initially, the only concern was the ability to maintain communication between devices and subsequently the error occurrence and packet loss during the process of communication. Later on, the need to take into attention issues such as congestion control of the network and the service differentiation [8]. The QoS parameters can be negotiated between the user and the network for each communication session, depending available resources and QoS requirements. However, providing a QoS guarantee implies that the performance of the network is consistent and predictable, which can be a major challenge [8].

3.2 Affecting Factors

The application layer QoS is affected by the width bandwidth, the delay, the delay variant, the error rate, the likelihood of loss, etc. For each parameters can be set the limits of variation. The factors that affect the QoS are related, not being able to improve a parameter without adversely change another [6]. Then, it is listed the most important parameters to obtain the QoS:

1. **Bandwidth:** For an application to run properly, it is needed a certain bandwidth Bluetooth connection, wherein the available bandwidth influences delays obtained data transfer. An application can explicitly specify the bandwidth needs or the bandwidth can be derived the application requirements for the delays. This one parameter is mostly determined by the algorithms polling performed by the master of the piconet [6].
2. **Delay and delay variation:** This parameter is caused by the available bandwidth and the retransmissions. The SCO traffic has priority over traffic ACL, so you can also contribute to delays in the ACL traffic. The delay-sensitive applications such as real-time applications and audio/video applications require few delays [6].
3. **Reliability:** All applications prefer that the data arrives in the correct order, however, some applications are more error tolerant than others. In general, applications audio/video can tolerate some errors, while application data, such as an HTTP request or email, require that the data arrives in the correct order. Increasing this parameter results in a decrease in bandwidth and hence an increase the delay [6].

4 Biomedical Applications

The main action of QoS in communication and networks is present among some factors that occur in the end systems, such as the delay end-to-end, the package loss, the variation in the delays as well as the throughput. The application of the QoS in these systems WLAN, as Bluetooth, can result in a higher efficiency of the system and a better applicability. Nowadays, Bluetooth can be applied in several biomedical applications, such as, sensors in hospitals and in patients' houses (live assistance 24 h/day). These mechanisms begin to have a higher relevance, because, with the population aging, an increase of diseases such as cardiovascular diseases, diabetes, and Alzheimer's disease, among many others, is expected. These diseases demand a certain supervision performed by the healthcare professionals and increase the costs associated with the healthcare [10, 17].

The use of devices or applications that rely on Bluetooth have much to offer to assist in this problem. Some examples of technologies that rely on the use of Bluetooth are described below. A biomedical application using Bluetooth, in order to have a higher control on a person's weight based om easy-of-use mechanism allows the self-control of the person's weight was developed [11]. The device created is composed by a monitor that implements an auto-monitoring system using a Bluetooth network, allowing a weight control. It is presented a prototype of the system but until now there were not performed tests in order to evaluate the system applicability as a personal health control unity. A low power A/D converter and it uses wireless networks, to make integration with other technologies or infrastructures easier and less expensive. This system is composed by two processors, the Bluetooth and the FPGA (Field-Programmable Gate Array), being these submersed in a solution. The developed system allows to convert an analogue signal that simulates an electrocardiogram (ECG) signal. This device works with a remote computer that processes data sent by Bluetooth. The test results with this system proved that it is possible to make a continual ECG signal transmission without information losses [12]. A biomedical application that monitors a person's health status in a daily basis. This application uses Bluetooth network that limits the system capacity to a small number of parameters. A prototype composed by a pulse clock sensor connected to a PDA (Personal Digital Assistant) was developed. The connection is made by Bluetooth. The system is also equipped with several components such as accelerometers and thermic sensors made by GSR (Galvanic Skin Reflex) electrodes. The Bluetooth module makes possible the connection between the PDA and a monitor allowing the health status evaluation of the person as well as is movements and behaviors. This way, the information acquired by the sensor allow to guide the person in real-time, giving him a self-control of its health [14]. A system for remote medical assistance of recovering infarct patients was also created [14]. These systems is implemented in some hospitals and it uses a Bluetooth connection to send data recorded by portable ECG recorders to mobile devices such as mobile phones. The data are then sent to a central unit implemented in a Web server where a database is implemented. The database organizes the information about the ECG and relevant clinical data about the patient, allowing a better access to the cardiologist. This way, it is possible to make a better motorization of the patient's health status.

5 Critical Analysis

Besides wireless networks, such as Bluetooth, having mobility as an advantage, sometimes this advantage introduces difficulties in the system that are not present in other networks. These difficulties are, for example, the quality of the wireless transmission, the great variability of the connection quality, the reduced bandwidth and the highly dynamic network topology, that allows the direction of the elements between the source and the end. Therefore, wireless networks require a more sophisticated QoS management. One of the requirements of this technology is to maintain the simplicity. Thus, the Bluetooth technology must continue to be a simple and low cost technology and the QoS must be as simple as possible, while responding to the applications needs. The QoS implementation should not result in a significant cost and energy consumption increase, because the demands to processing capacity, memory capacity and energy consumption must be limited. One of the disadvantages of this technology is that, if two slaves want communicate among themselves, they must do it through the master. Another disadvantage is related with the existence of delays and unavailability when the connection as interferences. It is important to mention that the QoS parameters should be defined as an extension of the actual Bluetooth specifications.

In the biomedical field, the implementation of the QoS results in a better efficiency of the Bluetooth, and consequently, increases the efficiency of the applications associated with this technology. For example, in the remote ECG it is demanded an almost null packet loss, because it is not plausible the presentation of a patient's ECG signal with losses. Bluetooth can be an important technology in the Pervasive Information fields and in the way of the systems communicates with ubiquitous devices.

6 Conclusions

In communication networks, the traditional vision of QoS is concerned with the end-to-end delay and its variation, packet loss and the provided bandwidth. Bluetooth devices communicate among themselves and form a network called a piconet, with one the master, and the others, the slaves. With the Bluetooth communications evolution, the users of this technology are becoming more demanding. Thus, it is necessary to develop mechanisms enabling the delivery of quality services that comply with the QoS parameters required by the users. Finally, this study provides a starting point for the construction and implementation of devices using Bluetooth technology in healthcare in order to increase the QoS in health institutions. It is an important asset to who wants to research in this area and did not have an idea how is application of Bluetooth in healthcare institutions.

Acknowledgments This work has been supported by FCT - Fundação para a Ciência e Tecnologia within the Project Scope UID/CEC/00319/2013.

References

1. U. Bilstrup, K. Sjöberg, B. Svensson, P.-A. Wiberg, A fault tolerance test enabling QoS in a Bluetooth piconet, pp. 33–36 (2004)
2. A. Abelha, et al., Agency for archive, integration and diffusion of medical information, in *Proceeding of AIDA* (2003)
3. L. Cardoso, F. Marins, F. Portela, M. Santos, A. Abelha, J. Machado, The next generation of interoperability agents in healthcare. Int. J. Environ. Res. Public Health **11**(5), 5349–5371 (2014)
4. H. Peixoto, M. Santos, A. Abelha, J. Machado, Intelligence in interoperability with AIDA, in *Foundations of Intelligent Systems SE—31*, vol. 7661, ed. by L. Chen, A. Felfernig, J. Liu, Z. Raś (Springer, Berlin Heidelberg, 2012), pp. 264–273
5. A. Ganz, Z. Ganz, K. Wongthavarawat, *Multimedia Wireless Networks: Technologies* (Standards and QoS, Pearson Education, 2003)
6. M. van der Zee, G. Heijenk, Quality of Service in Bluetooth networking—part I, no. TR-CTIT-01–01. University of Twente, Centre for Telematics and Information Technology (CTIT), Enschede, the Netherlands (2001)
7. A.B. Soares, Análise da Qualidade de Serviço de VPN—Redes Privadas Virtuais—Utilizando Redes Sem Fios (2014)
8. R.L. Aguiar, L. Brito, Qualidade de Serviço em redes móveis: presente e futuro, in *Conf Científica e Tecnológica em Engenharia—CCTE11* (2002)
9. T. Wu, C. Ke, C. Shieh, W. Hwang, A practical approach for providing QoS in bluetooth piconet, pp. 332–338
10. R. Latuske, A.R.S.S. Gmbh, Bluetooth health device profile and the IEEE 11073 medical device frame work bluetooth in medical applications, pp. 1–6
11. J. Parkka, M. van Gils, T. Tuomisto, R. Lappalainen, I. Korhonen, A wireless wellness monitor for personal weight management, in *Proceedings 2000 IEEE EMBS on information technology applications in biomedicine*, pp. 83–88 (2000)
12. J. Lönnblad, M. Ekström, A. Fard, J.G. Castaño, T. Johnson, Remote system for patient monitoring using bluetooth, in *IEEE International Conference on Sensors* (2002)
13. K. Ouchi, T. Suzuki, et al., LifeMinder: a wearable healthcare support system using user's context, in *Proceedings 22nd International Conference on Distributed Computing Systems Workshops*, pp. 791–792 (2002)
14. S. Khoor, J. Nieberl, K. Fugedi, E. Kail, Telemedicine ECG-telemetry with Bluetooth technology. Comput. Cardiol. **2001**, 585–588 (2001)
15. Volume 6 of the Bluetooth Core Specification, Version 4. Core System Package [Low Energy Controller Volume]. Issued on 17th December 2009

16. C. Gomez, J. Oller, J. Paradells, Overview and evaluation of bluetooth low energy: an emerging low-power wireless technology. Sensors **12**(9), 11734–11753 (2012)
17. A.H. Omre, S. Keeping, Bluetooth low energy: wireless connectivity for medical monitoring. J. Diab. Sci. Technol. **4**(2), 457–463 (2010)
18. Bluetooth low energy—White Peper—LitePoint, 5 Feb 2015. http://www.litepoint.com/whitepaper/Bluetooth%20Low%20Energy_WhitePaper.pdf
19. C. Bisdikian, An overview of the Bluetooth wireless technology. IEEE Commun. Mag. **39** (12), 86–94 (2001)

Computer Vision Based Indoor Navigation: A Visual Markers Evaluation

Gaetano C. La Delfa, V. Catania, S. Monteleone, Juan F. De Paz and J. Bajo

Abstract The massive diffusion of smartphones and the exponential rise of location based services (LBS) have made the problem of localization and navigation inside buildings one of the most important technological challenges of the last years. Indoor positioning systems have a huge market in the retail sector and contextual advertising; moreover, they can be fundamental to increase the quality of life for the citizens. Various approaches have been proposed in scientific literature. Recently, thanks to the high performances of the smartphones' cameras, marker-less and marked-based computer vision approaches have been investigated. In a previous paper, we proposed a technique for indoor navigation using both Bluetooth Low Energy (BLE) and a 2D visual markers system deployed into the floor. In this paper, we present a qualitative performance evaluation of three 2D visual markers suitable for real-time applications.

Keywords Indoor location systems · Computer vision · Fiducial markers

G.C. La Delfa (✉) · V. Catania · S. Monteleone
Department of Electrical, Electronics and Computer Engineering (DIEEI),
University of Catania, Catania, Italy
e-mail: gaetano.ladelfa@dieei.unict.it

J.F. De Paz
BISITE Research Group, Faculty of Science, University of Salamanca,
Plaza de la Merced S/n, Salamanca, Spain
e-mail: fcofds@usal.es

J. Bajo
Department of Artificial Intelligence, Polytechnic University of Madrid, Madrid, Spain
e-mail: jbajo@fi.upm.es

V. Catania
e-mail: vincenzo.catania@dieei.unict.it

S. Monteleone
e-mail: salvatore.monteleone@dieei.unict.it

A. Mohamed et al. (eds.), *Ambient Intelligence - Software and Applications*,
Advances in Intelligent Systems and Computing 376,
DOI 10.1007/978-3-319-19695-4_17

1 Introduction

The massive diffusion of the smartphone has contributed in the last years to create the conditions for a significant technological progress in the mobile consumer sector giving to developers and startups the perfect instrument to create innovative applications and services. Most of these applications and services are strictly related to the user position and context information: they are defined as Location Based Services (LBS) and are becoming very popular in the last years. Outdoor the GPS is almost a standard de facto for positioning and navigation, but in indoor environments does not actually exist a unique technology to solve the problem. Various approaches have been proposed in literature to address the challenge in a simple and scalable way, and also a lot of commercial solutions are appearing into the market. Among these, the most successful ones are those which take advantage of the hardware/sensors of the smartphone to extract contextual information and use them to localize the user. *Dead Reckoning systems* use accelerometer, magnetometer and gyroscope sensors embed into the smartphones to provide fast estimation of the user position [10]. *Received Signal Strength Indication systems* exploit the RSSI of the radio signals present in the environment (typically, WiFi signals, available for free in public buildings, or, recently, BLE signals) [6]. *Visible Light Communication systems* exploit the susceptibility of LEDs to the amplitude modulation at high frequencies to transmit information into the environment and perform accurate indoor positioning (without deteriorating the lighting functionality) [2, 8]. Recently, thanks to the high performances cameras and high computational capabilities of last generation smartphones, researchers are focusing on *Computer Vision systems* which rely on complex, (1) marker-less or (2) marker-based computer vision algorithms to determine the position of the user in the environment [1]. Usually, *Hybrid* techniques and technologies are used to improve the accuracy, reduce costs and enhance the performances of the whole indoor positioning system [14]. The remainder of this paper is organized as follows: in Sect. 2, we give an overview about the state of the art of the indoor localization approaches. In Sect. 3, we discuss about the features needed to build an efficient indoor localization and navigation system with 2D visual markers deployed onto the floor. Section 4 focus on analyzing three markers with respect to some parameters which are of interest for our use case. Finally we conclude with some considerations and ideas for future works.

2 Related Works

Nowadays, indoor navigation is a very hot topic and a lot of research has been made during the last decade. Researchers from Duke University proposed UNLOC [14] in which they merge environment sensing and dead reckoning (D.R.) to realize an indoor navigation system, based on the hypothesis that certain locations in indoor environments presents - in the sensors domains - identifiable signatures (Landmarks) generated by elevators, escalators, WiFi, etc. They use D.R. to track the user, and periodically reset the error when the user encounters a landmark. In [8], the authors suggest the use of LEDs and Visible Light Communications (VLC) to localize the user inside an environment in an accurate way. On the Tx side, the modification to the LED lighting infrastructure is cheap and simple, on the Rx side, Harald Haas [2] et al. show that it is possible to exploit the rolling shutter effect of CMOS-based cameras to let a mobile phone to decode the information transmitted by the LEDs infrastructure. Apple and Google included API for indoor positioning in their SDKs for iOS and Android. They use mixed technologies such as WiFi fingerprints, BLE, iBeacons and D.R. to perform the indoor localization task. The list of approaches, techniques and technologies is actually very long. Among all of these, as aforementioned, researchers are focusing on computer vision algorithms which use (1) marker-less approaches or (2) marker-based approaches for indoor localization purposes. Marker-less approaches are used when visual markers are undesirable due to aesthetic reasons; they rely on what the camera sees to deduce the position of the user and usually require a pre-knowledge of the environment. They are CPU-Intensive, and also require a considerable workload before they can start to work and frequent recalibrations. Marker-based approaches rely on 2D visual markers, which can be easily decoded even by a low-cost smartphone. B.L. Ecklbauer, in his thesis [3] uses the recognition of multiple custom ArtoolKit visual markers in a camera image to deduce the position of an Android smartphone, with no additional data sources, except the knowledge of the markers positions. In the report "An Indoor Navigation System For Smartphones" [1] the authors propose a simple color-based 2D visual marker, to obtain the user position, and orientation and a step detection algorithm, to track the user between two markers. Despite of the simplicity and scalability of all these techniques, there are some drawbacks such as (a) the need of a line of sight, (b) the sensitivity to light changes, (c) the size of the marker which must be as smaller as possible in order to be minimally invasive, (d) the fact that the app does not work in real-time, and (e) the cognitive workload for the user who has to look for the marker in order to auto-locate himself. In order to face all these drawbacks, in our previous paper [4] we proposed a hybrid approach which uses BLE for locating the user when there is not a line of sight and a 2D visual markers system deployed onto the floor (in order to let the user to auto-locate himself without any cognitive workload: in fact when he launches the app in navigation mode, the camera is in the palm of his hand and will be directed towards some part of the floor) to estimate the position with a good level of accuracy. To guarantee accuracy, minimal invasivity and real-time performances, it is important to choose the right marker according to the particular situation of deploy. In the previous paper we proposed the use of AruCo marker. In the present paper, we evaluate in a qualitative and empirical way the performances of three 2D visual markers: (a) Vuforia frame marker [7], ArUco marker [5], AprilTag [13].

3 Visual Markers Deployed onto the Floor: Requirements

To realize an efficient indoor navigation system using 2D visual markers deployed onto the floor, a critical point is the choice of the visual marker which best fits the particular place of deployment. Here we focus on some of the intrinsic features of the system, and on how we can exploit them to improve the performances:

- Almost uniform, prior-known background pattern of the floor (Fig. 1a). It is possible to use this feature to improve the speed of the decoding algorithm and to reduce the marker size.
- Almost fixed, prior-known size of the marker inside the frame (Fig. 1b). It depends on the distance between the camera - which is on the palm of the hand - and the floor, and makes easier and faster to find the marker in the frame, which brings to a detection speed improvement.
- Major probability for the marker to be in the upper part of the frame (Fig. 1c), due to the fact that when the user launches the app to navigate, he moves forward. It is possible to analyze a sub-portion of the frame and further improve the detection speed or use the saved time to apply some filters in order to enhance the quality of the image.
- Prior-known markers positions, so each decoded marker must be one in the boundary of the previously decoded marker.

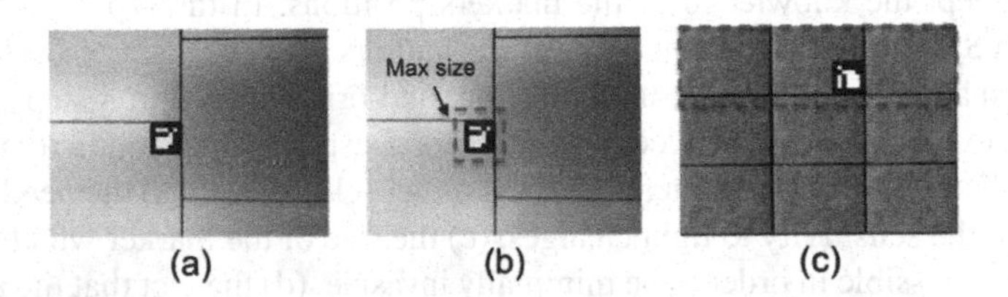

Fig. 1 (a) Uniform background pattern of the floor. (b) Max size of the marker inside the frame. (c) Major probability for the marker to be in the upper part of the frame

The characteristics that the chosen marker must have, considering that when the user use the system he is moving (usually with low speed), are:

1. *Small size*: it is required to reduce the invasivity of the system. We have to find the best compromise between size, speed of detection/decoding and robustness of the algorithm.
2. *Real-time detection*: To make the auto-localization process through visual markers transparent for the final user, the detection speed must be as fast as possible.
3. *Robustness to changes in light conditions*: it is required because typically the marker will be deployed in high dynamic environments characterized by the presence of other people, on/off switching of lights, shadows etc.
4. *Robustness in detecting blurred markers*: caused by too fast movement.

4 Real Time Visual Markers: Vuforia, ArUco, AprilTag

A visual marker system is composed by a set of 2D visual markers and a computer vision algorithm to detect and decode each marker using a smartphone camera or other computer vision technologies. Today, thanks to their low cost, flexibility and simplicity there are several visual markers in the market. Figure 2a shows the QR-Code: it can store up to 4296 characters, and contains a Reed-Solomon error correction algorithm which let to decode even partially occluded QR-Codes. Moreover, it is opensource and very well-documented but it has not real-time performances which make it not good for our purposes. Aztec code (Fig. 2b) is similar to the QR-Code (large amount of stored data, Reed-Solomon error correction algorithm) but it does not need a white border to be correctly decoded. To guarantee real-time performances, usually a visual marker which stores just a simple binary code is used. An example (Fig. 2c) is ArtoolKit marker [9]. Originally developed by Hirokazu Kato, the ArtoolKit library relies on a template matching algorithm to detect the marker. Thanks to that, the shape of an ARToolKit marker can theoretically be any image, surrounded by a black square. Other than the classical square markers we have also circular markers, stronger to perspective distortion and more precise, such as Intersense [12] (Fig. 2d). Figure 2e shows a marker invented by the MIT Media Lab [11], circular and with a diameter of just 3 mm. It can store a large amount of data, it is readable from 4 meters with a normal camera and it works by exploiting the Bokeh effect which occurs when the camera is out-of-focus. The analysis of the state of the art bring us to restrict the choice of the best marker that fits our requirements to three possible candidates (Fig. 3). We have chosen these markers also because they can be freely used through opensource, well-documented libraries or free SDKs and they are portable in all the major platforms. In the following, we give an overview of their features, strengths and weaknesses. We test the markers (with an iPhone 5S) in light, medium and dark floor pattern, in various light conditions.

Fig. 2 Visual markers examples

(a) Vuforia Marker

(b) AprilTag

(c) ArUco Marker

Fig. 3 Vuforia frame marker, ArUco marker, AprilTag

4.1 Vuforia Marker

Vuforia is an augmented reality multi-platform SDK developed by Qualcomm. It is very powerful and offers to the developers a lot of functionalities (objects, images, shapes and text recognition). Moreover, it can detect (and stimate the pose) of a special marker called Frame Marker (Fig. 3a), which we can use for our indoor localization purposes. There are 512 markers, distributed as an archive. Each one encodes an ID on the binary pattern along the border, and needs an area around it, free of graphical elements, with a good contrast respect to the black frame. The internal part of the marker is not used by the algorithm so it is possible to put inside an image, which makes it more good looking than other ones. We performed some detection/decoding tests for different sizes of the markers: (6.5×6.5), (5.0×5.0), (3.2×3.2) cm^2 and different distances marker-camera (80, 100, 120 cm), in movement and with the smartphone in the palm of the hand. We repeated the tests in several lighting conditions. Our tests show that a marker size of (6.5 × 6.5) and (5.0 × 5.0) cm^2, give good real-time performances for light, medium and dark floor pattern in good and average light conditions. The performances get worse for poor light conditions and if we reduce the size of the tag to (3.2×3.2) cm^2. Despite of the good performances the system has some drawbacks: (1) the source code is not accessible,so it is impossible to modify the algorithms in order to exploit the features of the floor, (2) the number of markers is fixed, which bring to a low flexibility, and (3) it is not possible to reduce a lot the size of the marker.

4.2 ArUco Marker

ArUco is a square visual marker realized by the AVA group from the University of Cordoba [5]. It can be decoded through the C++ ArUco library, which is cross platform (because openCV based), and opensource. It seems to be well-maintained by the research group. Differently from other similar systems, ArUco does not provide a predefined set of markers: it is possible to generate the desired number of markers, with the desired number of bits (n) encoded inside each of them. The library maximizes the inter-marker distance and the number of bit transitions and proposes an error correction algorithm which lets to correct a number of errors greater than the state of the art. It is also possible to estimate the pose of the marker with respect to the camera. Since ArUco does not have a fixed number of bits, the performances of the detection/decoding algorithm vary depending on this parameter, which can be set according to the requirements of our use case: small areas can be covered with few markers, which means that the n can be reduced, which in turn bring to a faster detection/decoding phase. We generated 512 ArUco markers and performed tests for the same size as the Vuforia marker tests, at the same distance marker-camera, under the same conditions. Our tests show that ArUco works very well, in any light conditions for a marker size of (6.5×6.5) and $(5.0 \times 5.0)\,\text{cm}^2$ with a distance marker-camera of $80cm$, 100 and 120 cm in any type of floor pattern. The performances are a little bit worse if we reduce the size of the marker to $(3.2 \times 3.2)\,\text{cm}^2$. ArUco source code is accessible for the developer: due to this, it is possible to modify the algorithms in order to adapt them to the scenario described in Sect. 3. Also the possibility to set the number of markers and bits increases a lot the flexibility of the system. In conclusion, ArUco is a good choice for an indoor localization system with visual markers deployed onto the floor, when the requirements are flexibility and real-time performances.

4.3 AprilTag

AprilTag is a square visual marker developed for robotic applications by Edwin Olson, at University of Michigan [13]. The opensource library lets to detect an AprilTag in an image, decode the ID of the marker, and stimate its 3D pose and orientation respect to the camera. The library is written in pure C with no external dependencies, and appears well-documented and well-maintained, robust to changes in light conditions and view angle, and with real-time performances. We performed some detection/decoding tests by choosing the recommended pre-generated markers family 36h11 (36 bit markers with minimum hamming distance between codes of 11), which consists of 518 different markers, and by using the same marker sizes, marker-camera distances and conditions of the previous Vuforia and ArUco cases. The results show that AprilTag works very well in all tested light conditions and for all tested sizes and marker-camera distances. The availability of the source code (which lets the developer to modify the algorithms to adapt them to the floor features), the speed of the system and the small marker size make AprilTag the best choice for an indoor, marker-based localization system when the flexibility about the number of markers is not required.

5 Conclusions and Future Works

In this paper, we have addressed the problem of choosing the best marker for an indoor navigation system with visual markers deployed onto the floor. We started with an overview on the state of the art, focusing on a marker-based computer vision approaches. Next we analyzed the particular use case of markers deployed onto the floor, in order to find some features that can be exploited to improve the performances of the system. The analysis lead us to choose three visual markers which have features and performances that match with our scenario: Vuforia marker, ArUco marker and AprilTag. We tested them, in different light conditions and floor patterns, by considering three main parameters: their size, their distance from camera and the detecting/decoding speed. The results show that Vuforia has good performances if the marker size is equal or greater than $(5.0 \times 5.0)\,\mathrm{cm}^2$, but the drawback is that its SDK is not opensource. AprilTag and ArUco have very good overall real-time performances in any tested light conditions and floor patterns, for all tested marker sizes. They are also opensource and cross-platform. While AprilTag seems to be a little bit quicker than ArUco and lets to reduce, more than ArUco, the size of the marker (while preserving overall performances), ArUco gives more flexibility because lets to generate the exact number of marker we require. We are planning to realize a proof of concept of our indoor localization system using both ArUco and AprilTag, in order to test better the approach in a real situations with both the markers, and to exploit the features of the floor. The goal is to reduce the size of marker (so the system can be less invasive) and enhance the speed. Moreover, we are investigating the possibility to mix this technique with D.R., to track the user between markers, and BLE localization.

References

1. A. Chandgadkar, W. Knottenbelt, *An Indoor Navigation System for Smartphones* (Imperial College, London, 2013)
2. C. Danakis, M.Z. Afgani, G. Povey, I. Underwood, H. Haas, Using a CMOS camera sensor for visible light communication, in *Workshops Proceedings of the Global Communications Conference, GLOBECOM 2012* (Anaheim, California, USA, 2012), pp. 1244–1248, http://dx.doi.org/10.1109/GLOCOMW.2012.6477759. Accessed 3–7 Dec 2012
3. B.L. Ecklbauer, A mobile positioning system for android based on visual markers. Ph.D. thesis, Hagenberg, Austria (2014)
4. G.C. La Delfa, V. Catania, Accurate indoor navigation using smartphone, bluetooth low energy and visual tags, in *Proceedings of the 2nd Conference on Mobile and Information Technologies in Medicine* (2014), http://mobmed.org/download/proceedings2014/mobileMed2014_paper_6.pdf
5. S. Garrido-Jurado, R. Muñoz-Salinas, F.J. Madrid-Cuevas, M.J. Marín-Jiménez, Automatic generation and detection of highly reliable fiducial markers under occlusion. Pattern Recognit. **47**(6), 2280–2292 (2014)
6. D. Han, S.H. Jung, M. Lee, G. Yoon, Building a practical wi-fi-based indoor navigation system. IEEE Pervasive Comput. **13**(2), 72–79 (2014). http://dx.doi.org/10.1109/MPRV.2014.24
7. Q. Incorporated, Qualcomm vuforia (2014), https://developer.vuforia.com/
8. A. Jovicic, J. Li, T. Richardson, Visible light communication: opportunities, challenges and the path to market. IEEE Commun. Mag. **51**(12), 26–32 (2013). http://dblp.uni-trier.de/db/journals/cm/cm51.html#JovicicLR13
9. H. Kato, M. Billinghurst, Marker tracking and hmd calibration for a video-based augmented reality conferencing system, in *IWAR '99: Proceedings of the 2Nd IEEE and ACM International Workshop on Augmented Reality* (IEEE Computer Society, Washington, DC, USA, 1999), pp. 85–94. http://dl.acm.org/citation.cfm?id=857202.858134

10. Y. Liu, M. Dashti, J. Zhang, Indoor localization on mobile phone platforms using embedded inertial sensors, in *WPNC* (IEEE, 2013), pp. 1–5. http://dblp.uni-trier.de/db/conf/wpnc/wpnc2013.html#LiuDZ13
11. A. Mohan, G. Woo, S. Hiura, Q. Smithwick, R. Raskar, Bokode: imperceptible visual tags for camera based interaction from a distance, in *SIGGRAPH '09: ACM SIGGRAPH 2009 Papers* (ACM, New York, NY, USA, 2009), pp. 98:1–98:8. http://doi.acm.org/10.1145/1576246.1531404
12. L. Naimark, E. Foxlin, Circular data matrix fiducial system and robust image processing for a wearable vision-inertial self-tracker, in *ISMAR '02: Proceedings of the 1st International Symposium on Mixed and Augmented Reality* (IEEE Computer Society, Washington, DC, USA, 2002), pp. 27–36. http://dl.acm.org/citation.cfm?id=850976.854961
13. E. Olson, AprilTag: a robust and flexible visual fiducial system, in *Proceedings of the IEEE International Conference on Robotics and Automation (ICRA)* (IEEE, 2011), pp. 3400–3407
14. H. Wang, S. Sen, A. Elgohary, M. Farid, M. Youssef, R.R. Choudhury, No need to war-drive: unsupervised indoor localization, in *Proceedings of the 10th International Conference on Mobile Systems, Applications, and Services (MobiSys '12).* (ACM, New York, NY, USA, 2012), pp. 197–210. http://doi.acm.org/10.1145/2307636.2307655

A Framework for the Secure Storage of Data Generated in the IoT

Ricardo Costa and António Pinto

Abstract The Internet of Things can be seen has a growing number of *things* that inter-operate using an Internet-based infrastructure and that has evolved during the last years with little concern for the privacy of its users, especially regarding how the collected data is stored. Technological measures ensuring users privacy must be established. In this paper we will present a technological framework for the secure storage of data. *Things* can then interact with the framework's API much in the same way they now interact with its current servers, after which, the framework will perform the required operations in order to secure the data before storing it. The methods adopted for the secure storage will maintain the sharing ability, conveniently allowing authorized access to other users, the initial user's terms (e.g. data anonymity) and the ability to revoke assigned privileges at all times.

Keywords Internet · Things · IoT · Secure · Storage · Database · Framework · API

1 Introduction

In the last years we have witnesses a unparalleled growth in the use of devices by human beings, making the, so called, Internet of Things (IoT) the most hyped technology in the planet [1, 2], and is expected to continue. Figure 1 clearly shows that the IoT is the emerging technologies that is most expected in a 5 to 10 years space. Such tremendous, completely uncontrolled, growth in such a short time frame has had no support from the traditional IT Industry. The market urged to respond to the

R. Costa (✉)
GCC, CIICESI, Escola Superior de Tecnologia e Gestão de Felgueiras,
Instituto Politécnico do Porto, Felgueiras, Portugal
e-mail: rcosta@estgf.ipp.pt

A. Pinto
GCC, CIICESI, Escola Superior de Tecnologia e Gestão de Felgueiras,
Instituto Politécnico do Porto, Portugal and INESC TEC, Porto, Portugal
e-mail: apinto@inesctec.pt

A. Mohamed et al. (eds.), *Ambient Intelligence - Software and Applications*,
Advances in Intelligent Systems and Computing 376,
DOI 10.1007/978-3-319-19695-4_18

consumers needs (or need to consume) and that has raised some complex problems, being the security and privacy some of the major ones, in our opinion.

1.1 IoT by the Numbers

According to the ABI Research [3], the installed base of active wireless connected devices will exceed 16 billion in 2014, about 20 % more than in 2013 and the number of devices will more than double from that to the 40.9 billion deviced expected for 2020. According to IDC, the IoT will represent a market value of $7.1 trillion in 2020, an impressive growth from the $1.9 trillion of 2013. Additionally, and just to name a few [4]:

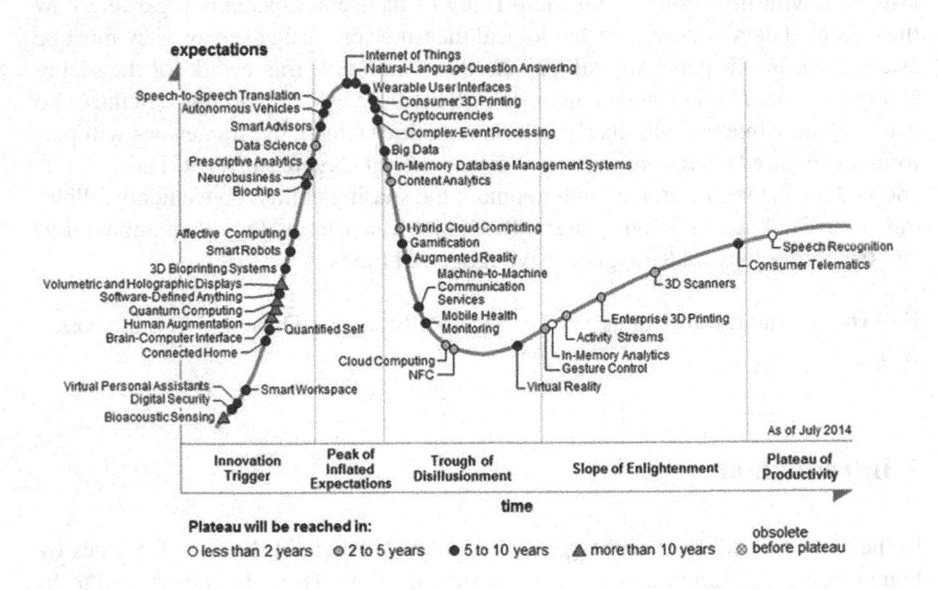

Fig. 1 Hype cycle for emerging technologies, 2014 [1]

- **IHS Automotive** says the worldwide number of cars connected to the Internet will grow to 152 million in 2020;
- **Navigant Research** says the worldwide installed base of smart meters will grow from 313 million in 2013 to nearly 1.1 billion in 2022;
- **On World** says the worldwide number of, Internet connected, wireless light bulbs and lamps will grow from 2.4 million in 2013 to over 100 million by 2020;
- **Juniper Research** says the wearables market will exceed the $1.5 billion in 2014, the double of its value in 2013.

Most of these devices, wearable or not, are trackers (or have the ability to track), or are health monitors, or record images and videos, among other functionalities. All of them, one way or another, collect data. This data is then sent to servers on the Internet that store it.

1.2 The Problem

The problem [5, 6], in our opinion, arises when data is put into the equation, specially if such data is considered sensible and private by its owner. Sensible and private data should not be stored in the traditionally way, due to its additional security and privacy requirements. This new kind of, massive collected, data is, however, being stored, without any special kind of security and privacy concerns, all around the world, in traditional databases (or new databases engines, but with the traditional store, search and access model) [6]. Due to the immense quantity of potential clients, effort has mainly been on server functionality, scalability, and responsiveness. Several solutions implement access control but store their data in clear text in their databases [7–13]. This data is, in many cases, private data that can even be potentially used to harm the original data owner (the only real owner, in our opinion) without is expressed permission. Some example scenarios of such potentially harmful situations are describes next.

Scenario 1 - Health Data: Assume a scenario of a user that practices sports, like running or biking, and uses his smart watch/phone/band to monitor his heart rate. He does so for a long period of time, years maybe, and then has a heart related health problem. The stored data has the potential to be used to determine the risk of that person suffering another heart related health problem health and, ultimately, to be used by insurance companies to not insure, or to stop insuring that person [14].

Scenario 2 - Track Data: Assume a scenario of a user, a truck driver, that has agreed to use a, remotely accessible, GPS-enabled device so that its employer can better estimate the delivery times, and supply those times to the company costumers. The stored data has the potential to be used by the employer to monitor and control its employee even outside the normal, day-to-day, work hours.

Scenario 3 - Pictures: Assume a scenario of a user that subscribes, on its mobile smartphone, a picture backup service that automatically sends the pictures to a cloud-based storage. In order to maintain the backup service scalability and its server CPU usage within working limits, the service does not encrypt the pictures in its databases, requiring only some form of user authentication to allow access. Such service can potentially be exploited, by means of a 0-day bug, for instance, and the stored pictures can be widely divulged in the Internet exposing the private life of the picture owners.

Scenario 4 - Contact information: Assume a scenario of a user that uses a cloud-based contact information backup service where he stores all his mobile phone contacts (names, cell phone numbers, street addresses, email addresses). The stored data has the potential to be sold by the backup service provider to advert companies or to aggressive phone selling companies.

This scenarios present some of the potential uses that such kind of data can have, even without the knowledge of its owner. Additionally, there are several companies that have huge profits for using, free of charge, such private data without paying any kind of compensation to the owners of the data.

2 Background and Related Work

The work related to ours can by categorized into the following two approaches: (1) IoT access standardization, and (2) data acquisition.
The first approach focus on tackling the heterogeneity found on the diversity of equipment, communication protocols and communication technologies that make the IoT and does so by creating a framework that makes them accessible in a uniform manner. In [7], the proposed service oriented architecture that hides the uniforms access devices via web services. It also supports service management, device management and security. In [8], another framework that tries to standardize IoT device access is proposed, in this case by making use of RESTful web services and with their services implemented as a OSGi (Open Service Gateway initiative) specification. More recently, in [9], another RESTful-based middleware framework was also proposed, differing by its use of the MQTT protocol [15]. Another RESTful middleware was also presented in [10], this one focusing more on the interaction with devices by means of the OpenMTC framework.
The second approach focuses on data acquisition, integration and storage in a way that still enables access control and the management of security and privacy. In [11], a platform for data acquisition and integration is proposed for IoT. This platform, in a first phase, collects data from devices and stores it in a cloud environment and, in a second phase, context-oriented mechanisms are executed over the data to produce context data. Personal data vaults, proposed in [12], is referred as a privacy architecture where user retain ownership of their data. Data is said to be filtered before sharing and user can take part in controlled data-sharing decisions. Their work is based on fine-grained Access Control Lists (ACL) and the capability to trace and audit operations logs. Neither solution use data encryption has a way of guarantying privacy nor consider user monetizing strategies.
Table 1 summarises the solutions that relate to our work. Each solution is identified by a name, is categorized by its approach and type, and analysed to verify if the solution implements access control mechanisms, if the data is encrypted prior to its storage, and if the user that owns the data is allowed to maintain control over the data sharing. For instance, the PDV (Personal data vault) solution falls in our data acquisition approach, is a web-based (HTTP) cloud-like solution that implements data access control, that allows the data owners to perform sharing control but stores its data in a clear text form.

Table 1 Related work

Name	Approach	Type	Access control	Data encryption	User control
SOCRADES	Access	Service-oriented	No	No	No
MAGIC Broker 2	Access	RESTful/OSGi	No	No	No
QEST	Access	RESTful/MQTT	No	No	No
M2M APIs	Access	RESTful	Yes	No	Unclear
Context	Data	OSGi	Yes	No	No
PDV	Data	Web-based	Yes	No	Yes

3 The Secure Framework for IoT (Sec4IoT)

In the traditional IoT architecture (Fig. 2) the *things* interact with the servers via an Internet-based infrastructure. Normally, the only security-related aspects that are considered are the ones related to the communication channels, often recurring to SSL/TLS protocols as the foolproof solution. We believe this is not enough since all the data that is stored, is stored in clear text form in their databases. Additionally, except [12], no solution gives their users the capability to either control data sharing, to audit data sharing or to revoke data sharing.

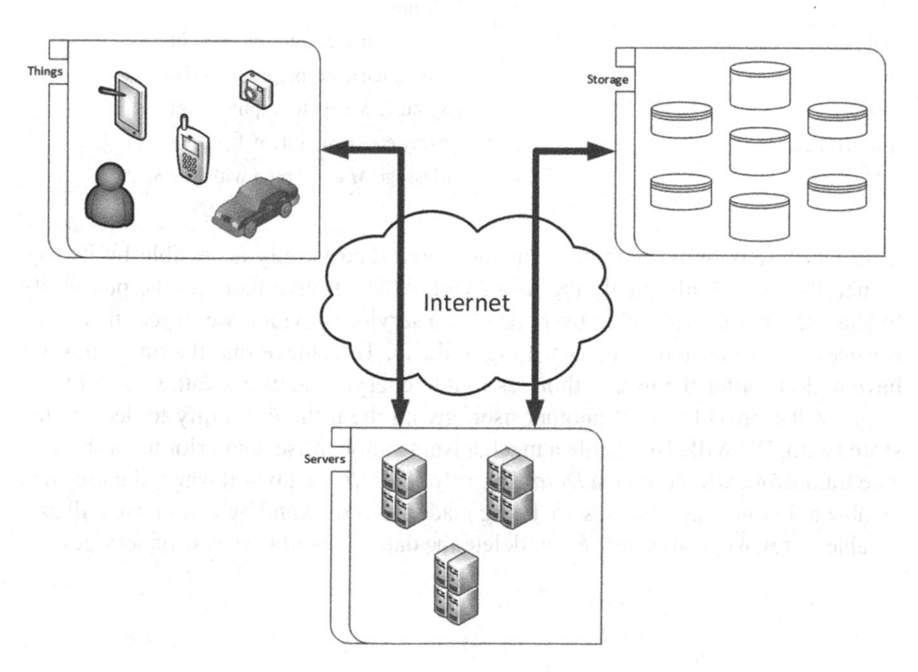

Fig. 2 IoT architecture

We propose adding an additional layer of security to the IoT tradition architecture (Fig. 3) and that such security layer be provided by our framework. Despite maintaining the traditional architecture and its communication channels, we require that the *things* use our framework API, named Secure Framework for IoT (Sec4IoT), for data sending and storing. The framework will implement an additional security layer providing, not only the needed user privacy, but also the user control of its own data. This additional security layer will be explained in the next set of steps, using the notation presented in Table 2:

1. **Initialization:** all the intervening parts of the process will need to pass by an initialization phase where each part will generate a public and private key pair [16];
2. The *thing* calls the API method for data upload - uploadData();
3. The uploadData() method will, before actually uploading any kind of data, generate a random secret key *secK*, encrypt the data with it and encrypt the *secK* with the user *pubK*, then it will generate a token *dataT* containing $[\{data\}_{secK}, \{secK\}_{pubK}]$ and, finally, upload the token to the servers;
4. The server will receive the encrypted data token dataT and store it in the database or datastore.

Table 2 Adopted notation

Notation	Meaning
pubK	Asymmetric cryptography public key
privK	Asymmetric cryptography private key
secK	Symmetric cryptography secret key
[item1, item 2]	Array containing item 1 and item 2
$\{M\}_K$	Message M encrypted with key K

This set of steps will guarantee that the stored data is only accessible by its true owner, the user. Additionally the framework API will give its users the possibility to share their data with other users or with a service provider, we expect that some services may require it, if he is willing to do so. To achieve that the only thing we have to do is, after the user authorizes it, it to encrypt the user's data *secK* with the *pubK* of the provider or of another user, giving them the capability to decrypt the stored data. We will also enable a mechanism to anonymise data prior to its sharing. The framework will contain a *Data Control and Audit* dashboard where the user will be able to know what accesses are being made to his data and by whom. He will also be able to revoke shares and, even, delete the data if possible (terms of service).

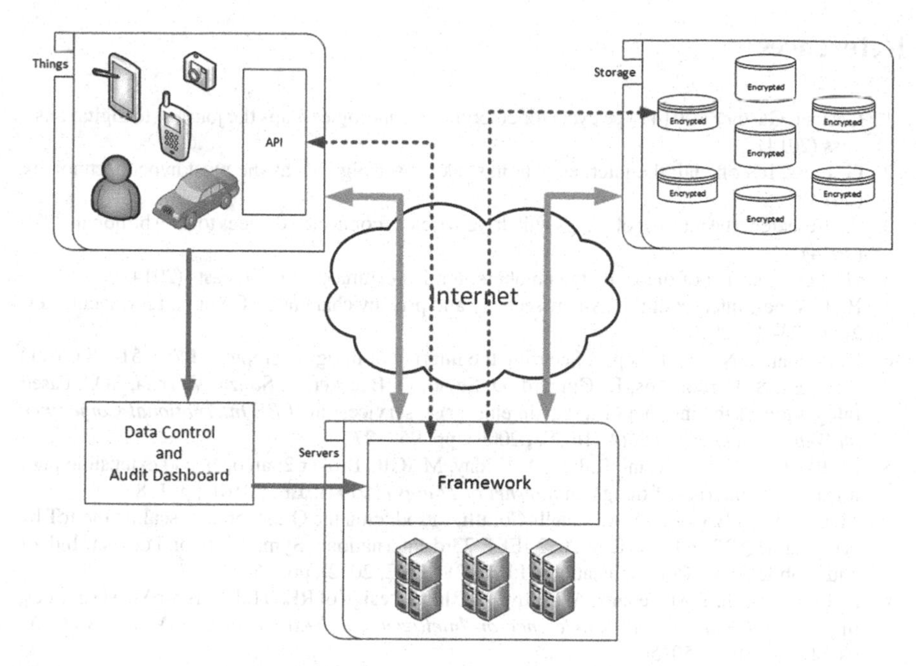

Fig. 3 IoT data storage new, framework, approach

4 Conclusions and Future Work

We have presented a new framework that is able to provide the additional and highly needed layer of privacy and security to the traditional IoT architecture. This additional layer enables users to regain their privacy rights, to collect their data only for their own consumption or to share it with a service provider or with other users, all accordingly to his own rules. We believe that this framework can be seemly introduced into the already in production solutions, turning them more secure from that moment on. Additionally, the proposed framework relegates the privacy control to the users, relieving services of such a burden. We also believe that this framework can also be easily adopted by new or ongoing developments, making them easier to include the needed security, and thus, accelerating its development and deployment. As future work, we plan to develop a proof-of-concept cloud-based data storage service that makes use of the proposed security framework and implements the mentioned data control and audit dashboard.

References

1. Gartner, Gartner's 2014 hype cycle for emerging technologies maps the journey to digital business (2014)
2. G. Press, It's official: the internet of things takes over big data as the most hyped technology (2014)
3. ABIresearch: the internet of things will drive wireless connected devices to 40.9 billion in 2020 (2014)
4. G. Press, Internet of things by the numbers: market estimates and forecasts (2014)
5. R.H. Weber, Internet of things-new security and privacy challenges. Comput. Law Secur. Rev. **26**(1), 23–30 (2010)
6. R. Roman, P. Najera, J. Lopez, Securing the internet of things. Computer **44**(9), 51–58 (2011)
7. P. Spiess, S. Karnouskos, D. Guinard, D. Savio, O. Baecker, L. Souza, V. Trifa, SOA-based integration of the internet of things in enterprise services, in *IEEE International Conference on Web Services ICWS 2009* (IEEE, , 2009), pp. 968–975
8. M. Blackstock, N. Kaviani, R. Lea, A. Friday, MAGIC Broker 2: an open and extensible platform for the internet of things, in *Internet of Things (IOT)* (IEEE, 2010), pp. 1–8
9. M. Collina, G.E. Corazza, A. Vanelli-Coralli, Introducing the QEST broker: scaling the IoT by bridging MQTT and REST, in 2012 IEEE 23rd International Symposium on Personal Indoor and Mobile Radio Communications (PIMRC) (IEEE, 2012), pp. 36–41
10. A. Elmangoush, T. Magedanz, A. Blotny, N. Blum, Design of RESTful APIs for M2M services, in *2012 16th International Conference on Intelligence in Next Generation Networks (ICIN)* (IEEE, 2012), pp. 50–56
11. Y.S. Chen, Y.R. Chen, Context-oriented data acquisition and integration platform for internet of things. in *2012 Conference on Technologies and Applications of Artificial Intelligence (TAAI)* (IEEE, 2012), pp. 103–108
12. M. Mun, S. Hao, N. Mishra, K. Shilton, J. Burke, D. Estrin, M. Hansen, R. Govindan, Personal data vaults: a locus of control for personal data streams, in *Proceedings of the 6th International COnference* (ACM, 2010), p. 17
13. D. Preuveneers, P. Novais, A survey of software engineering best practices for the development of smart applications in ambient intelligence. J. Ambient Intell. Smart Environ. **4**(3), 149–162 (2012)
14. P. Novais, R. Costa, D. Carneiro, J. Neves, Inter-organization cooperation for ambient assisted living. J. Ambient Intell. Smart Environ. **2**(2), 179–195 (2010)
15. OASIS Standard, MQTT Version 3.1.1. (2014)
16. Schneier, B.: Applied Cryptography: Protocols, Algorithms, and Source Code in C. 2 edition edn. Wiley (1995)

Multi-agent Systems for Classification of E-Nose Data

Yoshinori Ikeda, Sigeru Omatu, Pablo Chamoso, Alberto Pérez and Javier Bajo

Abstract Metal Oxide Semiconductor Gas Sensors are used to measure and classify odors. This kind of system requires both advanced sensor design and classification techniques. In this paper we present a MOGS (Metal Oxide Gas Sensor) specifically designed to classify the breath of humans. We propose an architecture that incorporates new sensing technology and a classification technique based on multi-agent systems. The proposal is evaluated using samples from Asian and European participants. The results obtained are promising.

Keywords Odor classification • Electronic nose • Multi-agent systems • Virtual organizations

1 Introduction

In recent years, odor sensor systems, also known as EN (Electronic Nose) systems have progressed significantly. The EN systems try to reproduce the sensation produced in humans when smelling certain odors. To do that, techniques like sensor arrays or pattern recognition systems are used [1].

Y. Ikeda (✉) · S. Omatu
Osaka Institute of Technology, Osaka, Japan
e-mail: ikeyoshi4871@gmail.com

S. Omatu
e-mail: omatu@rsh.oit.ac.jp

P. Chamoso · A. Pérez
Computer and Automation Department, University of Salamanca, Salamanca, Spain
e-mail: chamoso@usal.es

A. Pérez
e-mail: alberto.pgarcia@usal.es

J. Bajo
Department of Artificial Intelligence, Technical University of Madrid, Madrid, Spain
e-mail: jbajo@fi.upm.es

A. Mohamed et al. (eds.), *Ambient Intelligence - Software and Applications*,
Advances in Intelligent Systems and Computing 376,
DOI 10.1007/978-3-319-19695-4_19

In this document we are proposing an agent-based architecture for odor classification. These odors are taken with MOGSs, which are detailed in Sect. 4.

There are different kinds of MOGS, each one of them dedicated to certain specific odors. For instance, in [2] a fire detector is presented and, in [3], a tea classifier.

In our case, specifically designed sensors are going to be used to obtain human breath. To classify it, we will analyze data obtained by the sensors applying different classification techniques. Different specialized agents will apply these techniques and we will take as the final result the best one proposed by each agent [4].

2 Related Work

The "electro nose" term is generally associated to odor detection or the attempt to "smell" with technological devices. The most common technique is based on chemical gas sensors [5] and it will be used when developing this study. However, there are many advances in the developing of electro noses with other kind of sensors systems, such as optical sensor systems, mass spectrometry, ion mobility spectrometry, infrared spectroscopy, etc.

These systems have been used for different unalike tasks. For instance, in [6], a few applications of an EN system based on solid-state sensor arrays are presented to measure the air quality. In [7] ENs are applied to develop analysis in the food area, for example, to control the quality or to evaluate the freshness of products. More recently, and also related food odors, a system to differentiate between types of coffee has been presented in [8] and a system to develop a rapid diagnosis of enterobacteriaceae in vegetable soups in [9].

Nevertheless, during last recent years, the application of EN systems to analyze breath is increasing and important work with rats or even with human patients with chronic lung infections is being done [10] trying to detect the acute liver failure [11].

3 Measurement Method

In this paper, we have used a new device, shown in Fig. 1, to measure odor. This device is small size one and therefore, it is portable. Measurement system is shown Fig. 2.

Fig. 1 Measurement device

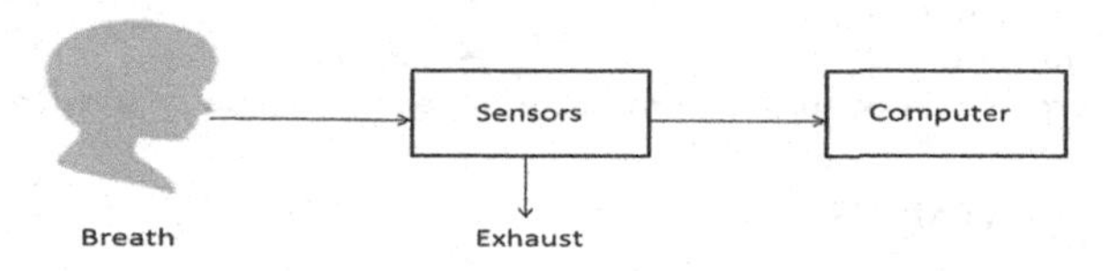

Fig. 2 Measurement system

Odor was measured by using the following procedure:

(i). Find out about subject's health.
(ii). Initiate the device.
(iii). Wait 10 s in order to stabilize it.
(iv). After that, measure subject's breath during 10 s.
(v). Finally finish measurement, stop the device and transmit the measurement results to the computer.

4 Principle of Metal Oxide Semiconductor Gas Sensors (MOGS)

We have used MOGSs to measure and classify odors. There are five kinds of sensor made by FIS Co. LTD (Japan). The characteristics of sensors are shown in Table 1. We have combined these sensors and classify various odors.

Table 1 Used sensors

Sensor number	Model number	Application
1	SB-AQS	VOC (volatile organic compound)
2	SB-15-00	Flammable gas (propane, butane)
3	SB-30-04	Alchol detection
4	SB-42A-00	Refrigerant gas (freon)
5	SB-31-02	Solvent detection

The working principle of MOGSs is explained in Fig. 3.

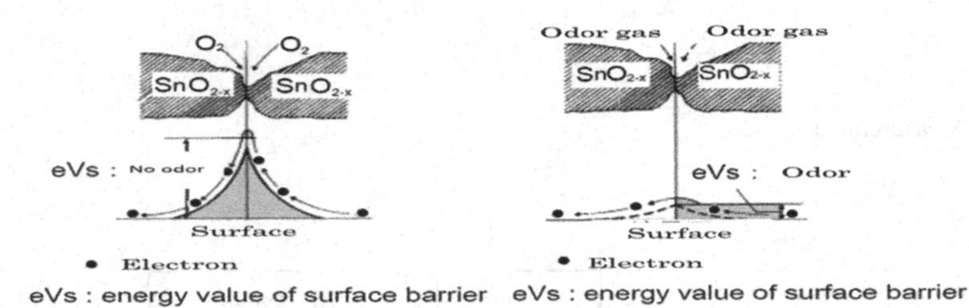

Fig. 3 The principle of MOGS

This kind of sensor makes good use of oxidation-reduction reactions. As shown in Fig. 3, the potential barrier changes attending to the existence or absence of gas. As a result, variable resistance becomes higher or lower.

Since the output voltage changes to low or high, we can measure odor information by using a measurement circuit as shown in Fig. 4.

Fig. 4 Odor measurement circuit

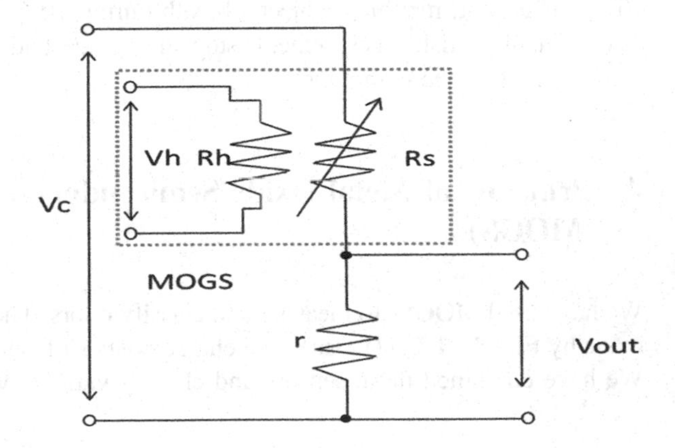

5 Classification Technique: Error Back-Propagation Type Neural Network (BPNN)

We have used the BPNN, as shown in Fig. 5, to classify odors. This classifier is one of the multi-layered neural networks. This method uses the error between an output value and a desired value to the given input. Connection weights are changed so that output error becomes the minimum based on the gradient method.

The error back-propagation algorithm is given by following steps.

Step 1: Set connection weights (W_{ji}, W_{kj}) to random numbers and set η (>0) as an initial value.
Step 2: Designate desired values of output $\{d_k, k=0, 1, \ldots, K\}$, corresponding to the input data $\{X_i, i=0, 1 \ldots, \mathrm{I}\}\}$ in the input layer.
Step 3: Calculate outputs in the hidden layer by the following formula:

$$net_j = \sum_{i=0}^{I} W_{ji}X_i - \theta_j, O_j = f(net_j), f(x) = \frac{1}{1+e^{-x}}$$

Step 4: Calculate outputs in the output layer by the following formula:

$$net_k = \sum_{j=0}^{J} W_{kj}O_j - \theta_k, O_k = f(net_k), f(x) = \frac{1}{1+e^{-x}}$$

Step 5: Calculate the error e_k and the generalization errors by the following formula:

$$e_k = d_k - O_k$$

$$\delta_k = e_k O_k(1 - O_k)$$

$$\delta_j = \sum_{k=1}^{K} W_{kj}\delta_k O_j(1 - O_j)$$

Step 6: Calculate half of root mean square error of all outputs by the following formula:

$$\mathrm{E} = \frac{1}{2}\sum_{k=1}^{K} e_k^2$$

Step 7: If E becomes a minimum, finish the learning, otherwise, connection weights should be changed by the following formula:

$$\Delta W_{kj} \equiv W_{kj}(t+1) - W_{kj}(t) = \eta\delta_k O_j$$

$$\Delta W_{ji} \equiv W_{ji}(t+1) - W_{ji}(t) = \eta\delta_j X_i$$

After changing the connection weights, go to Step 3.

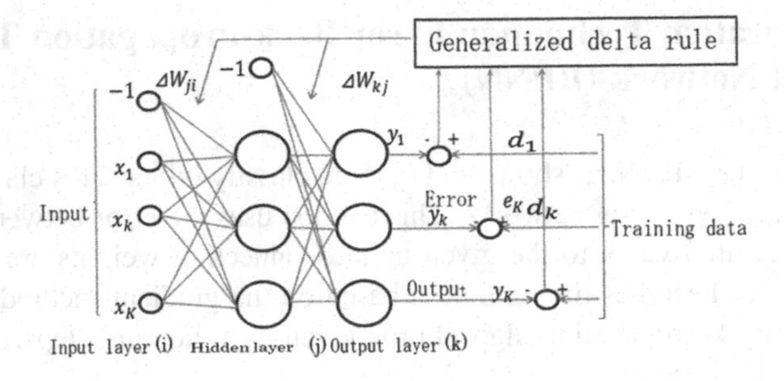

Fig. 5 BPNN

6 Case Study

6.1 Multi-agent Architecture

The proposed multi-agent architecture is based in a series of virtual organizations deployed in PANGEA platform [12]. This platform offers different tools for data base access, definition and creation of virtual organizations and integration with other platforms [13], which allow a simpler development. This architecture is based on the proposal from [4], with some modifications to adapt it to the specific case.

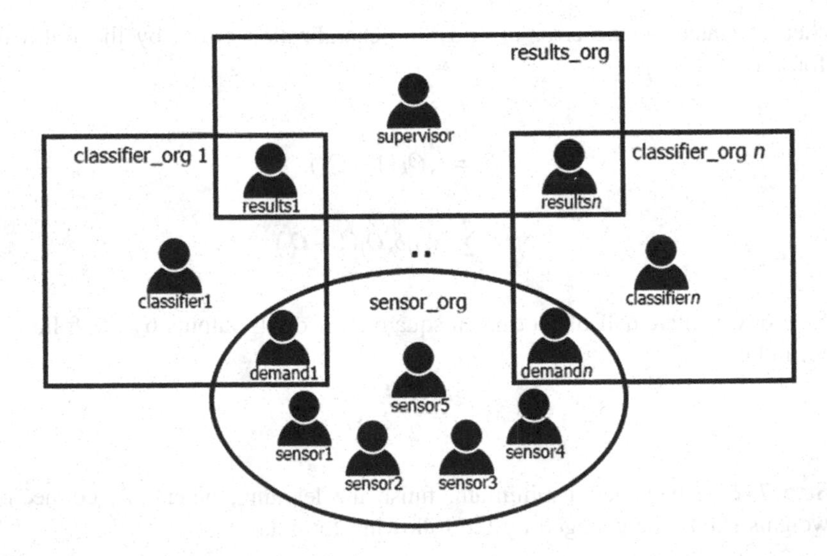

Fig. 6 Agent's organization scheme

Agents are structured in virtual organizations, with at least three of them necessarily and adding one by any other new classification method to be applied. Down below, the structure shown in Fig. 6 is described detailing each agent's functionality in the present system:

Sensor organization: there is an agent associated to each one of the 5 sensors (for this specific case), which will be the one in charge of getting the data offered by the sensor they represent and applying the necessary transformations to serve the data to the other organization participants. These participants are completed with the inclusion of a type of agent known as 'demand agent' and it will have an instance for every kind of classifier to be used. This agent is the one in charge of controlling that correct values are obtained from each one of the sensors participating in the analysis.

Classifier organization: it represents the virtual organization in which specialized agents are, in the application of the necessary techniques to develop classifications. To avoid the existence of a complex organization differing from the rest of organizations, the system has been structured in such a way that a virtual organization is generated by each kind of classifier. On each one of them, besides the necessary specialized agents to develop the classification, the 'demand agent' will have to exist as well as it exists on the 'sensor organization'. This time, its function will be to provide to the classifier agents the input data when they are ready. To receive classifiers output, the 'results agent' will also be required on each organization.

Results organization: on this organization we can find all the 'results agents' to provide data associated to each classifier output. The inclusion of a supervisor agent establishes the output provided to the final user according to the percentage indicated by the reliability grade on the response generated by the classifiers. In the case there are different results with the same reliability, the conflict gets resolved by returning the first response that gets to the supervisor agent. In any case, there will always have to exist a quicker-than-others result due to the fact that messages throughout PANGEA are delivered one by one.

6.2 Measurement of Odor Data

For this case study we have measured the breath of experimental participants from different precedence, precisely from Japan, China and Spain. When we measured the members' breath, we asked them about their health condition attending to a three-stage scale (good, usual, bad). These measurements were made several times to each subject and the average of the sample path of these data was used for each sensor. In Fig. 7, we show the measurement data of Japanese subjects. X-axis is the measurement of time and Y-axis is the output voltage.

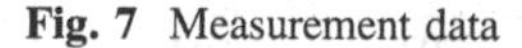

Fig. 7 Measurement data

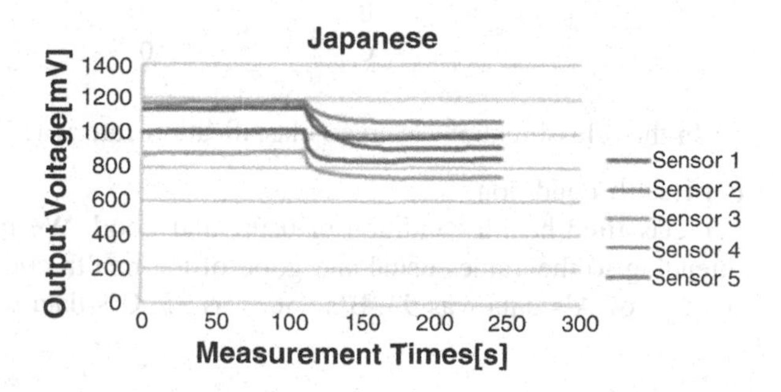

We used the minimum values from each sensor as the feature value and then, normalized these data. In this normalization, the maximum value is 1 and the minimum one is 0. We used neural networks as classification technique and then performed the classification.

7 Case Study

We classified the following case and showed conditions and results of these classifications. First, we explain about classification using by BPNN. η is set as 0.3.

After doing some classifications, we are showing the results and conclusions of them by explaining the classification using BPNN. η is set as 0.3.

(1) Japanese
We classified two persons. The total number for each data was 10. The training data and the test data are 5. The network was trained to learn until the error become less than, or equal to 1.0 × 〖10〗 ^(-3). Test data used 50 by changing parameters of neural networks and changing the test data and training data. The results are shown in Table 2.

Table 2 Classification results of Japanese (A, B)

Odor data	A	B	Total
A	26	24	50
B	29	21	50

In this classification, average classification rate was 46.5 %.

(2) Japanese, Chinese and Spaniard
We classified Japanese, Chinese and Spaniard. We also made a classification with subjects from Japan, China and Spain. The number of each data is four. The training data and the test data are 2. The network was trained to learn until the error become less than, or equal to 1.0×10^{-3}. Test data were 100 by changing parameters of neural networks and changing the test data and training data. The result is shown in Table 3.

Table 3 Classification result of Japanese (A), Chinese (B) and Spanish (C)

Odor data	A	B	C	Total
A	54	29	17	100
B	9	84	7	100
C	0	3	97	100

In this classification, average classification rate was 78.3 %.

(3) Health condition
We classified health condition of usual and good. We made another classification attending to the values usual and good of the health condition of the subjects. The number of A's data was 93, B's data was 35, C's data was 6 and D's data was 10.

The number of test data is 3 and that of training data is remaining data. The network performed learning until the error become less than, or equal to 5.0×10^{-3}. This classification performed 10 by changing parameters of neural networks and changing the test data and training data. The result is shown in Table 4.

Table 4 Classification result of usual (A, B, C) and good (D)

Odor data	A	B	C	D	Total
A	262	121	45	472	900
B	29	234	39	18	320
C	0	0	30	0	30
D	26	0	0	44	70

In this classification, average classification rate was 66.3 %.

In addition, we used Weka, which was developed by the University of Waikato, to classify cases (1) and (2). We used RBF network, SMO and Logistic as the classifiers to obtain the classification results. In case (1), SMO improved the average classification rate against the BPNN. The average classification rate was 68.4 %. In case (2), Logistic improved the average classification rate against the BPNN. The average classification rate was 90.9 %.

In case (2), classification rate was high. In case (1) and (3), classification rate was bad or usual. Therefore, we think that it isn't sufficient.

In order to improve classification results, we need to introduce a new classifier by using Weka. Regarding measurement condition, we think that it is ambiguous and isn't quantitative. Therefore, we need to think a better measurement condition.

References

1. K. El Hedhli, J.C. Chebat, M.J. Sirgy, Shopping well-being at the mall: construct, antecedents, and consequences. J. Bus. Res. **66**(7), 856–863 (2013)
2. J.E. Lane, Method, theory, and multi-agent artificial intelligence: creating computer models of complex social interaction. J. Cogn. Sci. Relig. **1**(2), 161–180 (2014)
3. Z. Fan, W. Duan, B. Chen, Y. Ge, X. Qiu, Study on the method of multi-agent generation algorithm within special artificial society scene. In Control (CONTROL), 2012 UKACC International Conference on (pp. 1076–1081). IEEE (2012)
4. P. Schreinemachers, T. Berger, An agent-based simulation model of human–environment interactions in agricultural systems. Environ. Model Softw. **26**(7), 845–859 (2011)
5. C. Zato, G. Villarrubia, A. Sánchez, J. Bajo, J.M. Corchado, PANGEA: a new platform for developing virtual organizations of agents. Int. J. Artif. Intell. **11**(A13), 93–102 (2013)
6. A. Mas, Agentes software y sistemas multiagente: conceptos, arquitecturas y aplicaciones (Prentice Hall, 2005)

7. G. Boella, J. Hulstijn, L. Van Der Torre, Virtual Organizations as Normative Multiagent Systems. In System Sciences, 2005. HICSS'05. Proceedings of the 38th Annual Hawaii International Conference on (pp. 192c–192c). IEEE (2005)
8. S. Rodriguez, V. Julián, J. Bajo, C. Carrascosa, V. Botti, J.M. Corchado, Agent-based virtual organization architecture. Eng. Appl. Artif. Intell. **24**(5), 895–910 (2011)
9. V. Julian, M. Rebollo, E. Argente, V. Botti, C. Carrascosa, A. Giret, Using THOMAS for Service Oriented Open MAS. In Service-Oriented Computing: Agents, Semantics, and Engineering (pp. 56–70). Springer Berlin Heidelberg (2009)

Using Machine Learning Techniques for the Automatic Detection of Arterial Wall Layers in Carotid Ultrasounds

Rosa-María Menchón-Lara, José-Luis Sancho-Gómez, Adrián Sánchez-Morales, Álvar Legaz-Aparicio, Juan Morales-Sánchez, Rafael Verdú-Monedero and Jorge Larrey-Ruiz

Abstract A fully automatic segmentation method for ultrasound images of the common carotid artery is proposed in this paper. The goal of this procedure is the detection of the arterial wall layers to assist in the evaluation of the Intima-Media Thickness (IMT), which is an early indicator of atherosclerosis and, therefore, of the cardiovascular risk. By measuring and monitoring the IMT, specialists are able to detect the incipient thickening of the arteries when the patient is still asymptomatic and to prescribe the appropriate preventive care. The proposed methodology is completely based on Machine Learning and it applies Auto-Encoders and Deep Learning to obtain abstract and efficient data representations. A set of 45 ultrasound images have been used in the validation of the suggested system. In particular, the resulting automatic contours for each image have been compared with four manual segmentations performed by two different observers. This study demonstrates the accuracy of our segmentation method, which achieves the correct recognition of the arterial layers in all the tested images in a totally user-independent and repeatable manner.

Keywords Machine learning · Deep learning · Auto-encoders · Ultrasound imaging · Intima-media thickness

1 Introduction

Cardiovascular diseases (CVD) remain the major cause of death in the world [10]. A large proportion of CVD are caused by an underlying pathological process known as atherosclerosis. Thus, its early diagnosis is critical for preventive purposes.

R.-M. Menchón-Lara (✉) · J. Sancho-Gómez · A. Sánchez-Morales ·
Á. Legaz-Aparicio · J. Morales-Sánchez · R. Verdú-Monedero · J. Larrey-Ruiz
Dpto. Tecnologí as de la Información y las Comunicacione, Universidad Politécnica de Cartagena, Plaza del Hospital, 1, 30202 Cartagena, Murcia, Spain
e-mail: rmml@alu.upct.es

J. Sancho-Gómez
e-mail: josel.sancho@upct.es

A. Mohamed et al. (eds.), *Ambient Intelligence - Software and Applications*,
Advances in Intelligent Systems and Computing 376,
DOI 10.1007/978-3-319-19695-4_20

Atherosclerosis involves a progressive thickening of the arterial walls by fat accumulation, which hinders blood flow and reduces the elasticity of the affected vessels.

The Intima-Media Thickness (IMT) of the Common Carotid Artery (CCA) is considered as an early and reliable indicator of atherosclerosis [9] and it is extracted from ultrasound scans [8], i.e. by means of a non-invasive technique. As can be seen in Fig. 1, blood vessels present three different layers, from innermost to outermost, intima, media and adventitia. The IMT is defined as the distance from the lumen-intima interface (LII) to the media-adventitia interface (MAI) . The use of different protocols and the variability between observers are recurrent problems in the IMT measurement procedure. To ensure the repeatability and reproducibility of the process, the IMT should be measured preferably on the far wall of the CCA within a region free of atherosclerotic lesions (plaques) [9], where a double-line pattern corresponding to the intima-media-adventitia layers can be clearly observed (see Fig. 1).

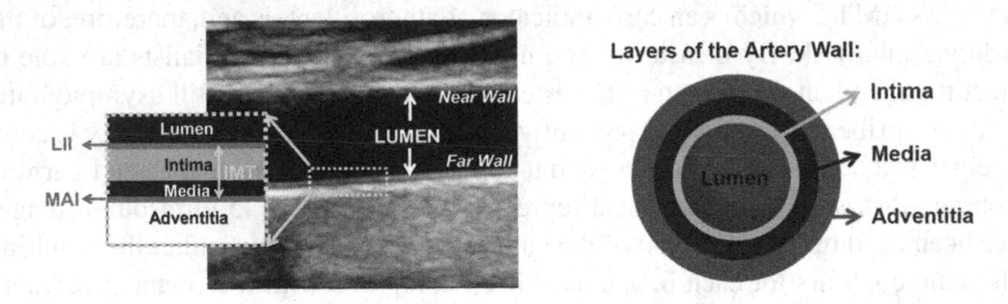

Fig. 1 Longitudinal view of the CCA in an ultrasound image (left) and diagram of the artery layers in a transverse section (right)

Usually, the IMT is manually measured by the specialist, who marks pairs of points on the LII and MAI. It is possible to reduce the subjectivity and variability of manual approaches and to detect the IMT throughout the artery length by means of image segmentation algorithms. This work addresses an automatic segmentation technique completely based on Machine Learning to jointly extract the LII and MAI from ultrasound CCA images and, therefore, to assist in the reproducibility and objectivity of the IMT evaluation.

The remainder of this paper is structured as follows: Sect. 2.1 describes the dataset of ultrasound CCA images, while Sect. 2.2 introduces the machine learning concepts used in this work. In Sects. 2.3 and 2.4, the proposed segmentation method is explained in detail. The accuracy of the results obtained by our automatic segmentation is analysed in Sect. 3. Finally, the main extracted conclusions close the paper.

2 Methodology

Figure 2 shows an overview of the proposed segmentation methodology. Firstly, a given ultrasound CCA image is pre-processed to automatically detect the region of interest (ROI), which is the far wall of the blood vessel. Then, those pixels belonging to the ROI are classified to detect the LII and MAI. As commented in Sect. 1, our method is completely based on Machine Learning. In particular, the concepts of Deep Learning and Auto-Encoders (AE) have been included in these stages.

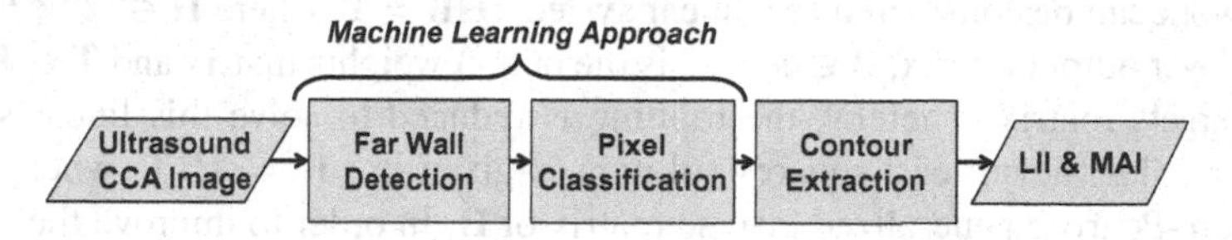

Fig. 2 Flow chart of the proposed method for the CCA segmentation

2.1 Image Database and Manual Delineations

The set of images used in this work consists of 45 ultrasounds of the CCA taken with a Philips iU22 Ultrasound System at different scale resolution (ranging from 0.033 to 0.081 mm/pixel). All of them were provided by the Radiology Department of Hospital Universitario Virgen de la Arrixaca (Murcia, Spain).

To assess the performance of the proposed segmentation method, it is necessary to compare the automatic results with some indication of reference values. In our case, the *ground-truth* corresponds to the average of four manual segmentations for each ultrasound image. In particular, two different observers delineated each image twice, with a mean period of one month between tracings. Each manual segmentation of a given ultrasound image includes tracings for the LII and MAI on the far carotid wall. The delineations were performed by marking at least 10 points over the images for each contour, which were subsequently interpolated. Hereinafter, we will refer to the different segmentations as: MA1 and MA2 for manual segmentations from observer A; MB1 and MB2 for manual segmentations from observer B; GT for the ground-truth and AUT for our automatic segmentation.

2.2 Machine Learning Approach

In the last decade, Extreme Learning Machine (ELM) has emerged as a powerful tool in the learning process of Single-Layer Feed-Forward Networks (SLFN) by providing good generalization capability at fast learning speed [4]. Given N arbitrary distinct samples $(\mathbf{x}_n, \mathbf{t}_n)$, where $\mathbf{x}_n \in \mathbb{R}^d$ (input vectors) and $\mathbf{t}_n \in \mathbb{R}^m$ (target vectors), the output of a SLFN with L hidden neurons (Fig. 3 (left)) and activation function $f(\cdot)$ is given by

$$\mathbf{y}_n = \sum_{j=1}^{L} \beta_j f(\mathbf{w}_j \mathbf{x}_n + b_j), n = 1, ..., N; \quad (1)$$

where $\mathbf{w}_j$ is the input weight vector connecting the input units and the j-th hidden neuron, β_j is the output weight vector connecting the j-th hidden neuron and the output units, and b_j is the bias of the j-th hidden neuron.

ELM is based on the randomly initialization of the input weights of SLFN. Thus, the network can be considered as a linear system: $\mathbf{HB} = \mathbf{T}$, where $\mathbf{H} \in \mathbb{R}^{N \times L}$ is the hidden layer output matrix, $\mathbf{B} \in \mathbb{R}^{L \times m}$ is the output weights matrix and $\mathbf{T} \in \mathbb{R}^{N \times m}$ is the targets matrix. Thereby, the training is reduced to solve this linear system whose smallest norm least-squares solution is given by: $\hat{\mathbf{B}} = \mathbf{H}^{\dagger}\mathbf{T}$, where $\mathbf{H}^{\dagger}$ is the Moore-Penrose generalized inverse matrix of $\mathbf{H}$. In order to improve the robustness and generalization performance, a regularization term (C) can be added to the solution [3]:

$$\hat{\mathbf{B}} = \left(\frac{\mathbf{I}}{C} + \mathbf{H}^T\mathbf{H}\right)^{-1} \mathbf{H}^T\mathbf{T} \tag{2}$$

Although ELM provides an efficient training for SLFN, the performance of machine learning methods and applications highly depends on the selected features for the representation of the problem. Thus, to make progress towards the Artificial Intelligence (AI), the new perspectives in Machine Learning are necessary based on learning data representations that make more accurate classifiers/predictors [1]. In this sense, Deep Learning has emerged as set of algorithms that attempt to model more abstract and more useful representation of the data by means of architectures with multiple non-linear transformations [2]. Among the various deep learning architectures, this work focuses on deep networks based on Auto-Encoders (AE). In particular, the ELM auto-encoders (ELM-AE) introduced in [5] have been used to solve our segmentation task. Auto-encoders are SLFN (Fig. 3 (left)) performing unsupervised learning in the sense that an AE is trained to reconstruct its own inputs, i.e. $\mathbf{t}_n = \mathbf{x}_n$. Therefore, in the hidden layer of the AE takes place a feature mapping: if $L < d$ (number of hidden neurons < input data dimension), a compressed data coding is obtained as hidden layer output ($\mathbf{H}$); while if $L > d$, the result is a sparse representation of data.

2.3 Far Wall Detection

This section describes the first stage of the proposed methodology, in which the carotid far wall is located in a completely automatic way by means of a deep architecture for Pattern Recognition (see Fig. 3 (right)). An ELM-AE has been designed to obtain useful and efficient representations of image blocks for their posterior classification as *'ROI-block'*, if the pattern of the far wall is recognized, or *'non-ROI-block'*, otherwise. The size of the image blocks to process is 39 × 39 pixels, i.e. the ELM-AE has an input data dimension of 1,521 features. The optimal coding is obtained with 850 hidden neurons. Therefore, a compressed representation of the image block is obtained. The connections between these new features ($\mathbf{H}$) and the output (y) are analytically calculated according to Eq. (2). The dataset used in the learning process of this system consists of 13,776 patterns (50 % from each class): two thirds for training and the remaining for testing. The proposed architecture has shown good performance with a classification accuracy of 98.85 ± 0.08 % over the test samples. Furthermore, the correct detection of the far wall is achieved in all the 45 tested images, even in noisy and blurred ones and when the artery appears tilted or curved in the image because of the probe position or the own anatomy of the subject. Some examples can be seen in Fig. 4.

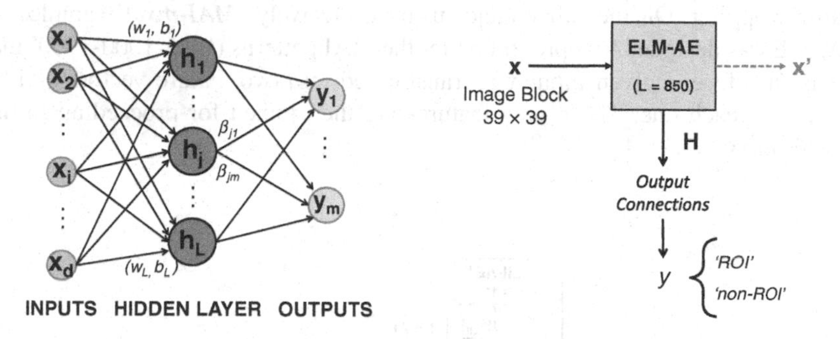

Fig. 3 Structure of a generic SLFN (left) and overview of the strategy for the far wall detection in CCA ultrasounds (right)

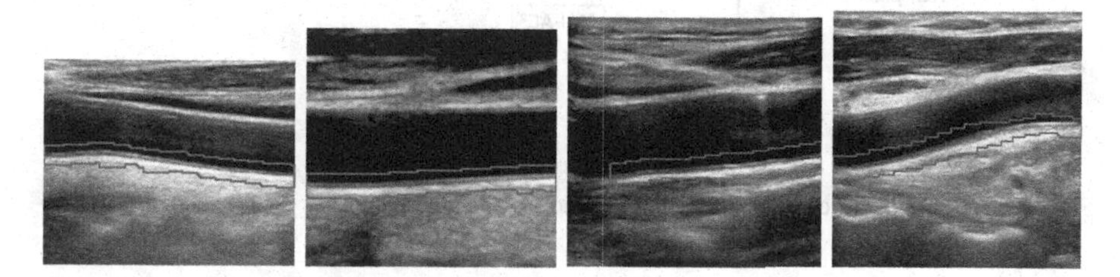

Fig. 4 Examples of CCA far wall detection

2.4 Arterial Layers Segmentation

The segmentation of the LII and MAI in the ultrasound images is carried out by means of a classification of pixels belonging to the ROI, which is an idea that has already been applied in previous works of the authors [6, 7]. In particular, the intensity values from a certain neighbourhood centred on the pixel to classify provide the necessary contextual information to the classifier for the recognition of the arterial layers. In the present study, the neighbourhoods consist of 51 × 5 pixels (255 input features) and four classes have been considered as outputs: *'LII-pixel'*,

'MAI-pixel', *'med-pixel'* (media layer) and *'non-IMC'* (out of the intima-media complex). Figure 5 shows the proposed deep network. The dataset used in the design and training process of this architecture consists of 38,908 patterns (two thirds for training and the remaining for testing), which are distributed equally among the four classes.

With the aim of obtaining meaningful representations from the inputs corresponding to LII and MAI, two different ELM-AE have been implemented. On one hand, the LII-AE was trained (layer-wise unsupervised training) only with samples of class *'LII-pixel'* and it consists of two stacked stages, which perform a '255-1100-1900' feature mapping. On the other hand, using exclusively *'MAI-pixel'* samples, the MAI-AE was designed to represent better the MAI patterns ('255-1000-1900' mapping). Therefore, a given input **x** is transformed into two feature vectors of 1,900 dimensions each one. These new features are then joined for proceeding to their classification.

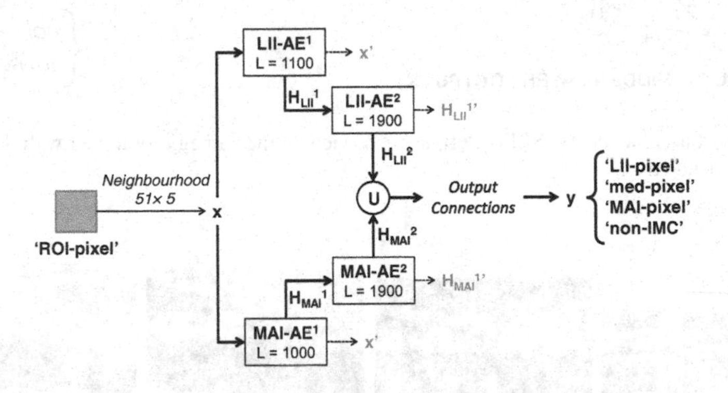

Fig. 5 Deep-architecture designed for the LII and MAI segmentation

The proposed system achieves a high success rate: 99.76 ± 0.03 % of accuracy in the recognition of LII patterns is obtained, while the classification accuracy for MAI samples is 99.69 ± 0.04 %. Figure 6 shows an example of the automatic segmentation results. As can be seen in the central picture, thc IMT boundaries are properly identified and the classification results cover the variability of the manual segmentations (see the superimposed points). However, it is necessary to define the final contours from these results. Thus, two curves are fitted to the peaks of intensity gradient within the LLI and MAI-pixels, respectively (Fig. 6 (bottom)).

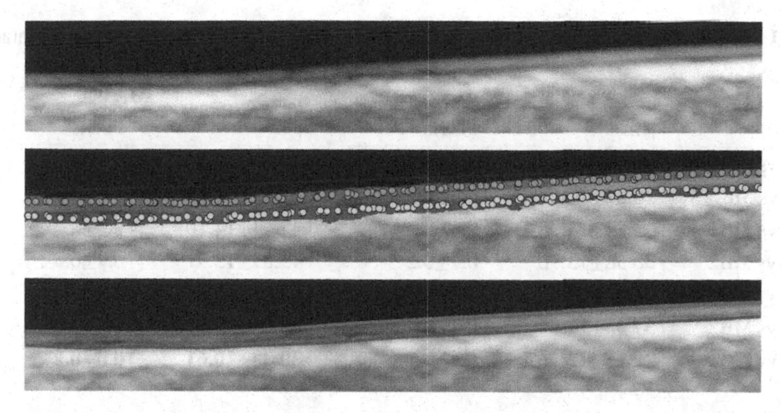

Fig. 6 Example of automatic segmentation results. Far wall detected in the raw image (top); classification results and points from manual segmentations (central): 'LII-pixels' in red and 'MAI-pixels' in blue; final LII and MAI contours (bottom)

3 Results

As commented above, in order to validate the precision of the segmentation results, the automatic contours have been compared with four manual tracings performed by two different experts. On the one hand, the intra-observer error is assessed in each case by comparing two manual segmentations from the same expert (MA1 vs MA2 and MB1 vs MB2). Furthermore, the inter-observer errors of the manual segmentations are also evaluated (MA1 vs MB1, MA1 vs MB2, MA2 vs MB1 and MA2 vs MB2). These comparisons characterize the uncertainty and variability of the manual procedure. On the other hand, the inter-method error is evaluated by comparing our automatic segmentations with those considered as ground-truth (AUT vs GT). The different segmentation errors are calculated separately for LII and MAI using the Mean Absolute Distance (MAD). This metric is based on the vertical distance between contours along the longitudinal axis of an image.

Figure 7 shows the distributions of the segmentation errors for LII and MAI over the 45 tested ultrasound images. In each box, the 25th percentile, the median and the 75th percentile of the corresponding error are emphasised. This statistical analysis reveals that a greater variability exists for the MAI, which is even more noticeable between manual segmentations. This is due to the fact that, in general, transitions from lumen to intima layer are clearer than transitions from media to adventitia layer. Moreover, Table 1 shows the mean and standard deviation of the different errors over our image database. Since the scale resolution varies from one image to another, the results in terms of pixel are also included for a better characterization of the difference between segmentations. In view of the results, it is possible to appreciate that our automatic segmentation reduces the uncertainty and variability of the manual procedure and, therefore, it will lead to a more reliable and precise measurement of the IMT.

Table 1 Difference (MAD metric, mean and standard deviation) between different segmentations

	LII		**MAI**	
	μm	Pixels	μm	Pixels
MA1 vs MA2	35.14±12.03	0.7±0.3	45.00±18.35	0.7±0.4
MB1 vs MB2	28.62± 8.05	0.5±0.2	41.49±11.91	0.8±0.3
MA1 vs MB1	33.74±14.38	0.6±0.2	45.44±14.10	0.9±0.2
MA1 vs MB2	36.56±13.70	0.7±0.2	45.89±13.91	0.9±0.3
MA2 vs MB1	36.89±13.44	0.7±0.3	49.86±17.44	1 ± 0.4
MA2 vs MB2	37.33±13.94	0.7±0.2	48.84±15.99	1 ± 0.4
AUT vs GT	**22.15±12.16**	**0.4±0.2**	**30.40±10.61**	**0.6±0.2**

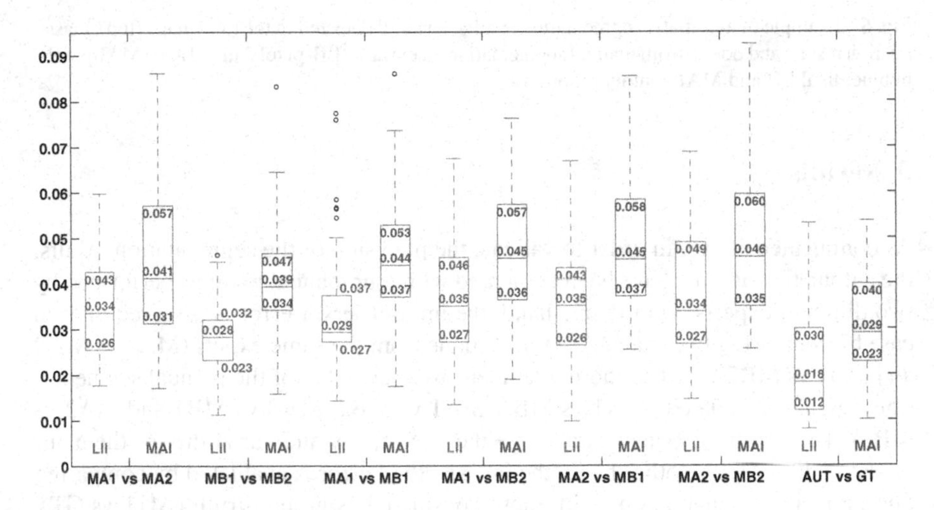

Fig. 7 Statistical analysis of mean absolute distance (mm) between different segmentations

4 Conclusions

This paper proposes a segmentation method for ultrasound images of the common carotid artery completely based on Machine Learning, in order to detect the arterial far wall and to extract the lumen-intima and media-adventitia interfaces in a reliable and automatic way. In particular, the suggested architecture is based on the Extreme Learning Machine. Furthermore, the concepts of Deep Learning and Auto-Encoders have been used to obtain useful data representations for solving the segmentation task, which is posed as a pattern recognition problem.

The method has been tested over a database of 45 images and its accuracy is assessed by comparing the automatic contours with four manual segmentations performed by two different observers. The results show a mean segmentation error of 0.022 mm for the lumen-intima interface and 0.030 mm for the media-adventitia interface, that represent scarcely half a pixel. Besides, our automatic segmentation avoids the subjectivity of manual procedures. Thus, it is possible to conclude that it will lead to a reproducible and more reliable measurement of the intima-media thickness.

Acknowledgements Authors would like to thank the Radiology Department of 'Hospital Universitario Virgen de la Arrixaca' (Murcia, Spain) for their kind collaboration and for providing all the ultrasound images used.

References

1. Y. Bengio, A. Courville, P. Vincent, Representation learning: a review and new perspectives. IEEE Trans. Pattern Anal. Mach. Intell. **35**(8), 1798–1828 (2013)
2. L. Deng, D. Yu, Deep learning: methods and applications. Technical Report MSR-TR-2014-21 (January 2014), http://research.microsoft.com/apps/pubs/default.aspx?id=209355
3. G.B. Huang, H. Zhou, X. Ding, R. Zhang, Extreme learning machine for regression and multiclass classification. IEEE Trans. Syst. Man Cybern. Part B-Cybern. **42**(2), 513–529 (2012)
4. G.B. Huang, Q.Y. Zhu, C.K. Siew, Extreme learning machine: theory and applications. Neurocomputing **70**(1–3), 489–501 (2006)
5. L.L.C. Kasun, H. Zhou, G.B. Huang, C.M. Vong, Representational learning with extreme learning machine for big data. IEEE Intell. Syst. **28**(6), 31–34 (2013)
6. R.M. Menchón-Lara, M.C. Bastida-Jumilla, J. Morales-Sánchez, J.L. Sancho-Gómez, Automatic detection of the intima-media thickness in ultrasound images of the common carotid artery using neural networks. Med. Biol. Eng. Comput. **52**(2), 169–181 (2014)
7. R.M. Menchón-Lara, J.L. Sancho-Gómez, Fully automatic segmentation of ultrasound common carotid artery images based on machine learning. Neurocomputing **151**(Part 1(0)), 161–167 (2015)
8. K.S. Nikita, Atherosclerosis: the evolving role of vascular image analysis. Comput. Med. Imaging Graph. **37**(1), 1–3 (2013)
9. P.J. Touboul et al., Mannheim carotid intima-media thickness and plaque consensus (2004–2006-2011). Cerebrovasc. Dis. **34**, 290–296 (2012)
10. WHO, Global atlas on cardiovascular disease prevention and control. Online, www.who.int/cardiovascular_diseases/en/

eServices - Service Platform for Pervasive Elderly Care

Isabel Marcelino, Rosalía Laza, Patrício Domingues, Silvana Gómez-Meire and António Pereira

Abstract In this paper, we present a solution to improve elderly's quality of life. eServices – Service platform for pervasive elderly care was designed to aggregate several services developed to meet senior population's needs. It monitors basic life signs, environment variables and uses personal location technology. Besides sensor services, eServices solution contains digital services align with emotional and social care needs. Due to target population specifications, eServices was designed to be as simple and accessible as possible in order to remove technological barriers. One of eServices major features is to detect imminent danger situations, act accordingly. It also collects the data from the sensors, location routines and from the interactions between the elderlies and the provided services to detect behavior deviations in order to act preventively. The platform was tested by seniors in real scenario. The experimental results demonstrated that the proposed platform was well accepted and easy to use by seniors, which demonstrated enthusiasm and interest in daily bases use.

Keywords Pervasive healthcare • Gerontechnology • Older adults • Smart environment • Health platform

I. Marcelino (✉) · R. Laza · S. Gómez-Meire
Higher Technical School of Computer Engineering, University of Vigo,
Polytechnic Building Campus Universitario as Lagoas s/n, 32004 Ourense, Spain
e-mail: isabel.marcelino@ipleiria.pt

R. Laza
e-mail: rlaza@vigo.es

S. Gómez-Meire
e-mail: sgmeire@vigo.es

I. Marcelino · A. Pereira
INOV INESC Innovation, Institute of New Technologies of Leiria, 2411-901 Leiria, Portugal
e-mail: apereira@ipleiria.pt

I. Marcelino · P. Domingues · A. Pereira
School of Technology and Management, Computer Science and Communications Research Centre, Polytechnic Institute of Leiria, 2411-901 Leiria, Portugal
e-mail: patricio.domingues@ipleiria.pt

A. Mohamed et al. (eds.), *Ambient Intelligence - Software and Applications*,
Advances in Intelligent Systems and Computing 376,
DOI 10.1007/978-3-319-19695-4_21

1 Introduction

Demographics analysis shows that world population is aging [1]. This factor creates new challenges in order to maintain sustainability. As elderlies are more willing to have health issues it is vital to follow and monitor their status, not only to react in distress situations, but also preventively. Additionally, emotional and social problems also deserve the same care as physical forum.

Along with world's population aging factor, active workers in modern societies are facing time poverty. Family (children, parents), friends and loved ones, demanding jobs, participation in community tasks, bureaucratic responsibilities (taxes, insurance, bills), plus commuting and keeping up with endless 24/7 online communications (social networks, emails) leaves young adults with the feeling of failure in several of these aspects of life because there just isn't enough time for everything. One of these aspects is precisely to give proper care to their elderly family members [2].

To overcome these issues, several concepts and technologies can be applied to improve elderly's quality of life: gerontechnology. Gerontechnology concept was defined as bringing technologies to deal with age-related problems [3]. For instance wireless sensor networks (WSN) to monitor the environment; body area networks (BAN) to gather basic life signs; personnel locator technologies like Radio-Frequency IDentification (RFID) and GPS (Global Positioning System) to grant that help is sent exactly where it's needed; ambient assisted living (AAL) to assure that elderlies stay autonomous as long as possible in their favorite environment (usually in their own homes); human-computer interaction (HCI) to exceed digital exclusion in senior population; cloud computing to eliminate IT obstacles, allowing IT environment more flexible and scalable; pervasive healthcare either implantable, wearable, portable and environmental, in both rural and urban areas.

In this paper we present a solution that combines all of the stated concepts and technologies, providing a service ecosystem to senior population, called eServices - Service platform for pervasive elderly care.

The remainder of the paper proceeds as follows: related work is presented in Sect. 2, followed by the presentation of eServices in Sect. 3. Section 4 contains the evaluation and results. Conclusions are presented in Sect. 5.

2 Related Work

Many solutions are being developed with the purpose to improve elderly's quality of life using emerging technologies.

Some of them gather bio information, like VitaDock [4]. Another commercial example is Microsoft Health Vault [5] that stores user's health data, have apps and services to analyze and manage that information and to share it. Microsoft also has, in its innovation to society topics, the issue "Gaming can promote mobility, balance and coordination in the ageing population" using devices such as Kinect for Xbox 360 for that purpose [6]. Also in the commercial field, there are several companies developing sensors to gather vital parameter information. Some of them use external sensors, others textile sensors, or even implanted sensors [7]. Other companies are developing alarm services, like panic buttons in bracelets, necklace or mobile phones, associating them with telecare service [8]. Other improve senior well-being by using emotion regulation techniques [9]. There are also European and other states initiatives aligned with this motivation. Its refer in [10] six European projects towards a future where people can have an active role managing their own healthcare, especially seniors: CAALYX, eCAALYX, COGKNOW, EasyLine+, I2HOME, and SHARE-it. Other examples are AAL4ALL [11] or UniversAAL [12]. Fall detection issue is focused in [13] solution.

Despite innumerous and valid initiatives, some issues are yet to solve, such as: the complexity of the proposed solutions, the sometimes lack of involving end users in the development process and ear their needs, base alarm detection situations only in physical parameters monitoring leading to false positives, only react to situations without preventing them. To overcome the issues identified, we propose eServices. Our solution will be discussed in detail in the next section.

3 eServices - Service Platform for Pervasive Elderly Care

eServices is a novel platform with a ecosystem of services targeted for senior population that stands for the remaining solutions due to its unique group of features:

- Regarding elderly's needs as a whole, not only the physical aspect, but also the emotional and social part as health comprehended not only physical parameter monitoring, but also the promotion of psychological well-being;
- Empowering simplicity were the actions seniors need to perform occur in a natural way, integrated and not interfering with their daily routine, intuitively, without memory efforts. We think that simplicity is the key to a successful solution given the target population;
- Having in mind a solution to embrace technologically illiterate or even unlettered individuals;
- Promoting user-centered design approach to suppress barriers between elderlies and technology usage;
- Granting a reactive and preventive solution that will be able to detect behavior deviations as accurate as possible due to the several input conjunction: basic vital signs from biosensor monitoring, personnel locators and routine agents (when and where is usually the user), interaction between the user and service usage (what services are used, in what schedule);
- Enabling security and removing ethical issues due to sensitive information by avoiding quasi-identifiers and protect user identities as long as cyphering the information;

- Encapsulating in a middleware platform all the services available for elderlies, allowing a unique central access, removing integrity issues and the complexity of having several separate pieces;
- Maintaining a scalable architecture that allows the transparent addition of end users, service providers and even the copulation of more sending features in sensor nodes;
- Endorsing a pervasive solution to provide its availability anytime, anywhere, any device, any access, any people, any service,…, AnyN;
- Removing any installation to use services available in web applications;
- Thinking about the costs supported by senior population. Many elderlies cannot afford complex and complete solutions as a whole. It may seem a better proposal to have a modular solution with de basic and additional modules to append as needed and affordable.

To ensure the mentioned group of features, eServices architecture comprehends 3 major modules: end users clients, service providers and middleware platform, as present in Fig. 1. Service providers will register their service in the service catalogue present in the middleware platform, while end users will access the services they are authorized.

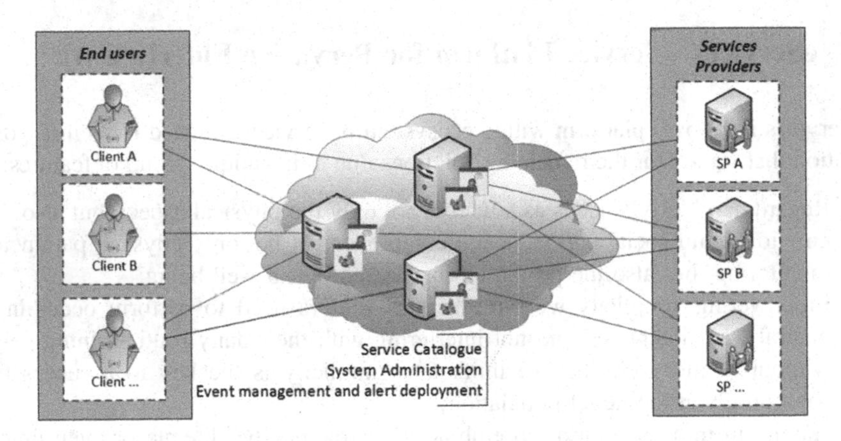

Fig. 1 eServices architecture

3.1 End User Client

This module is composed of several specifications: biosensors, smart home, personnel locators and service access device.

Biosensors: small and non-invasive BAN sensors will monitor basic life signs aggregating and sending them over Global System for Mobile Communications (GSM) (in outdoor scenario) or local network (indoor). Our research team as developed a more specific work in fall detection since, in our interviews made to medical staff, they appoint out fall detection as one of the major distress signal regarding health problems in older adults [14–16].

BAN will be used for reactive and preventive response. On one hand to immediately detect urgent situations and trigger a proper response. On other hand, to send information periodically in order to build profiles. The collected information will allow to specify what ranges of values are normal for a specific individual as long as feed pattern recognition system to anticipate or early detect risk situations;

Smart home: while the elderly is in his habitation (home, nursing home) there are environment sensors that will record events such as fire, gas, movement detection, burglar and intruder alarms, etc. and activate a proper response;

Personnel locators: with GPS (outdoor) or RFID (indoor). This will let, not only to send specialized help in alarm situations, but also to obtain routine information that will be an input to behavior deviation patterns;

Service access device: a device (computer, tablet) will be present in the senior habitat. It will serve 2 purposes: to aggregate information from sensors and send it over network as long as to provide interaction between the elderly and the remaining services. Regarding the web application that will allow the elderly to interact with the different services, inclusive design, gerontodesign, simplicity and user experience are our major concerns. The interaction between the elderly and the services will also be an input to behavior deviation patterns as it is possible to check the consumed services, the usual times of usage, etc. [17].

3.2 Middleware Platform

The middleware platform module will be the bridge between end users and service providers. This scalable solution will allow providing services on demand. Users will be able to request services that are not available to consume.

This module is composed of several specifications: service catalogue, system administration and event management.

Service catalogue will provide a range of services to end users, while service providers will register their service in the catalogue. Some categories were already identified: medical, maintenance, call center, leisure and cultural.

In medical category, users will be able to have virtual doctor appointments, using a web cam. The idea is not to replace face-to-face consults, but to add this possibility for those who have less mobility and also to reduce expenses in outings to health centers. Medical category also integrates the possibility to schedule appointments and exams. It also integrates the reminders to medication.

Maintenance category may be used to request repairman's such as plumbing, electricity, requesting medication household deliveries, help in tax payments, ask for home hairdresser, shopping's or other small errands. These small errands can also be archived by Neighbor-to-Neighbor Time Bank approach. This approach allows people to realize some volunteering work and also trade services. Another available service is to integrate ads and contracts to rent rooms by older people to students near universities, creating a symbiotic relationship. The students may have low cost or free rooms and seniors (especially the ones that live alone) may have company and help in daily activities.

Call center category is developed for supporting users either to respond to doubts over services usage, either to have on-line psychologists to provide specialized help.

Leisure category provides access to streaming TV, music streaming, games and other leisure activities. Games include some to improve and memory stimulation and others to infer and correct if elderly's posture is deteriorating as long as to enhance physical exercise. The leisure category will also provide a service focused on the elderly to allow musical expressiveness through motion, solely using the resources available in an ordinary home computer [18].

Cultural category includes virtual visits to monuments and e-learning classes. This category will also include experiences records, recording memories and knowledge, including handcrafts, recipes, folk medicine, proverbs, traditional agriculture methods, etc. It is extremely important to guarantee that senior tacit knowledge get in life experience can persist across later generations.

Along with the service catalogue, the system administration will also be present in middleware platform module and will be responsible for the platform administration tasks across a back office. It will include issues such as user and service administration, security, authentication, authorization, access control, accountings, ciphering information, managing quasi-identifiers, data replication and balanced distribution methods through the cloud.

Event management and alert deployment will also be present in middleware platform module and will be responsible for the data collection of all the platform and detect alert situations as long as to predict potential risk situations. It will also trigger the correct response, either by sending specialized help or to notify family members, according to previous specifications.

3.3 Services Provider

Service providers will have to register their service in one of the existing categories or ask to create a new one. Service providers will also need to use middleware platform communication protocol. After the registration, icons and voice commands need to be created in order to grant accessibility from end users that will subscribe the service.

4 Evaluation and Results

Regarding evaluation and the methodology applied, we have develop a prototype of eServices and implemented 3 services: one to make a video call to the user's doctor (in medical category), a Portuguese card game called "sueca" (in leisure category), a service to allow sending an asynchronous voice message (in call center category) [19].

A qualitative research method by: (1) a brief explanation and demonstration of the system; (2) acceptance tests were the individuals where asked to fulfil 3 tasks (one in each service), along by our observation to monitor tentative and successful task accomplishment, user errors, times to perform each action when compared to younger and with technological skills; (3) semi-structured interviews with control questions to obtain user's profiles and to acquire feedback about satisfaction, further usage and importance in daily use.

About population sample, acceptance tests were made to 11 seniors with 81 average ages. 5 males (with 63, 76, 84, 87 and 91 years old) and 6 females (with 73, 80, 80, 83, 84 and 87 years old). All of them are technologically illiterate; none of them have any computer skills or interaction. Two of them are unlettered. To compare the times to perform the task between older adults and younger adults, another subset was also subject of analysis. The sample consisted of 7 young adults with 28 average ages. 2 males (with 37 and 54 years old) and 5 females (with 34, 34, 34, 35, 35 years old). All with technological skills (Table 1).

Table 1 Tasks

T1	Make a video call to the doctor
T2	Play a full "sueca" game against computer
T3	Send an asynchronous voice message

It can be observed in Fig. 2 that older adults are taking the same amount of time to completing the tasks when comparing with younger adults. The exception is in task 2, where it is asked to perform a card game. After a careful analysis we have concluded that the extra time was not to use the provided interface, but in playing the game itself. Therefore, we can infer that elderlies are profiting of the services without any age restriction or technological barrier. From the conducted semi-structured interviews, all of the elderlies presented joy in experience eServices and transmit that it was something that they could use in daily bases to improve their quality of life.

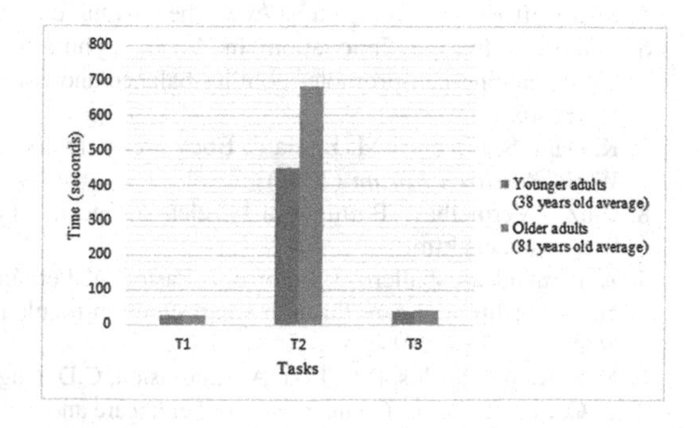

Fig. 2 Average time (in seconds) in completing each task, by younger adults and older adults

5 Conclusions

As previously mentioned, the world population is aging. Senior population are more likely to depend on others since health problems starts to appear, many of them live by themselves experience loneliness and insecurity feelings, they stop sense to play an active role in society. Moreover, the speeded-up life that younger adults are facing does not allow them to give a proper care to their elderly family members. Besides, the inversion tendency between active and retired people, is leading governments to have massive pension expenses as long as medical care expenses due to impaired health status usually presented by elderlies.

Several solutions are being developed by exploring emergent technologies we present to overcome these issues. In this paper we present the solution eServices that stands out due to a group of unique features. eServices architecture comprehends end client users, service providers and a middleware platform. Besides presenting stating eServices features and its architecture, we have developed a prototype of eServices and 3 services that were furtherly evaluated by acceptance tests and semi-structured interviews. eServices is an asset to improve elderly's quality of life and welfare, developing secure feelings not only in elderlies, but also in their family members as they are reassured that proper response is triggered in imminent danger situations. eServices also provides tools to potentiate digital and social inclusion in senior population.

Acknowledgments This work was partially funded by the [14VI05] Contract-Programme from the University of Vigo.

References

1. U. Nations, Population ageing and development 2012 (2012)
2. U. Nations, UNITED NATIONS report of the expert group meeting 'family policy in a changing world: promoting social protection and intergenerational solidarity (2009)
3. H. Bouma, J. Graafmans, A framework on technology and aging. *Gerontechnology* **3** (1992)
4. VitaDock, http://www.vitadock.com/vitadock.html
5. Microsoft Health Vault, https://www.healthvault.com/pt/en
6. Microsoft Europe Innovation in Society, http://www.microsoft.eu/innovation-in-society/videos/gaming-can-promote-mobility-balance-and-coordination-in-the-ageing-population-cm1l.aspx
7. K. Dam, S. Pitchers, M. Barnard, Body area networks: towards a wearable future, in *Wireless World Research Forum* (2001)
8. Cruz Vermelha Portuguesa Teleassistência, http://www.cruzvermelha.pt/actividades/teleassistencia.html
9. A. Fernández-caballero, J. Latorre, J. Pastor, A. Fernández-Sotos, Improvement of the elderly quality of life and care through smart emotion regulation. Ambient Assist. Living Daily Act. **8868**, 348–355 (2014)
10. M.N. Kamel Boulos, R.C. Lou, A. Anastasiou, C.D. Nugent, J. Alexandersson, G. Zimmermann, U. Cortes, R. Casas, Connectivity for healthcare and well-being management: examples from six European projects. Int. J. Environ. Res. Public Health **6**(7), 1947–2071 (2009)

11. L. Camarinha-Matos, J. Rosas, A. Oliveira, F. Ferrada, A collaborative services ecosystem for ambient assisted living, in *13th IFIP WG 5.5 Working Conference on Virtual Enterprises (PRO-VE)* (2012)
12. S. Hanke, C. Mayer, O. Hoeftberger, H. Boos, R. Wichert, M.-R. Tazari, P. Wolf, F. Furfari, universAAL – an open and consolidated AAL platform, in *Ambient Assisted Living* 2011, pp. 127–140
13. J. Castillo, D. Carneiro, J. Serrano-Cuerda, P. Novais, A. Fernández-Caballero, J. Neves, A multi-modal approach for activity classification and fall detection. Int. J. Syst. Sci. **45**(4), 810–824 (2014)
14. F. Felisberto, N. Moreira, I. Marcelino, F. Fdez-Riverola, A. Pereira, Elder care's fall detection system, in *Ambient Intelligence Software and Applications* 2011, pp. 85–92
15. F. Felisberto, N. Costa, F. Fdez-Riverola, A. Pereira, Unobstructive Body Area Networks (BAN) for efficient movement monitoring. Sensors **12**(9), 12473–12488 (2012)
16. F. Felisberto, F. Fdez-Riverola, A. Pereira, A ubiquitous and low-cost solution for movement monitoring and accident detection based on sensor fusion. Sensors (Basel) **14**(5), 8961–8983 (2014)
17. I. Marcelino, D. Lopes, M. Reis, F. Silva, R. Laza, A. Pereira, Using the eServices platform for detecting behavior patterns deviation in the elderly assisted living : a case study. Biomed. Res. Int. (2014)
18. L.S. Reis, G. Reis, J. Barroso, A. Pereira. AMIGA - an interactive musical environment for gerontechnology, in *4th International Conference on Software Development for Enhancing Accessibility and Fighting Info-exclusion (DSAI 2012)* 2012, vol. 14, no. Dsai, pp. 208–217
19. A. Pereira, A. Leal, L. Reis, I. Marcelino, A. Gaspar, J. Patrício, Patent - Método de comunicação assíncrono de mensagens de voz e audio, PT 10796 (2012)

A Sensor-Based Framework to Support Clinicians in Dementia Assessment: The Results of a Pilot Study

Anastasios Karakostas, Georgios Meditskos, Thanos G. Stavropoulos, Ioannis Kompatsiaris and Magda Tsolaki

Abstract This paper presents the main mechanisms of a sensor-based framework to support clinical diagnosis of people suffering from Alzheimer disease and dementia. The framework monitors patients at a lab environment while trying to accomplish specific tasks. Different types of sensors are used for monitoring the patients, while a graphical user interface enables the clinicians to access and visualize the results. Sensor data is semantically integrated and analyzed using knowledge-driven interpretation techniques based on Semantic Web technologies. Moreover, this paper presents encouraging preliminary results of a pilot study in which 59 patients (29 Alzheimer disease –AD– and 30 mild cognitive impairment –MCI) participated in a clinical protocol. Their analysis indicated that MCI patients outperformed AD patients in specific tasks of the protocol, verifying the initial clinical assessment.

Keywords Alzheimer · Sensors · Semantic interpretation · Daily activities

---

A. Karakostas (✉) · G. Meditskos · T.G. Stavropoulos · I. Kompatsiaris · M. Tsolaki
Centre for Research and Technology Hellas, Information Technologies Institute,
6th Km Charilaou-Thermi Road, 57001 Thermi, Greece
e-mail: akarakos@iti.gr

G. Meditskos
e-mail: gmeditsk@iti.gr

T.G. Stavropoulos
e-mail: athstavr@iti.gr

I. Kompatsiaris
e-mail: ikom@iti.gr

M. Tsolaki
e-mail: tsolakim@iti.gr

A. Mohamed et al. (eds.), *Ambient Intelligence - Software and Applications*,
Advances in Intelligent Systems and Computing 376,
DOI 10.1007/978-3-319-19695-4_22

1 Introduction

The frequency of dementia is rising all over the world, with considerable socio-economic impacts that are creating an imperative need for finding effective means of treatment. In current clinical practice, treatment of dementia begins with its diagnosis, which is based on behavioral assessments and cognitive tests that highlight quantitative and qualitative changes in cognitive functions, behaviors and activities of daily life, characteristic of the dementia syndrome and its underlying diseases. Typical questionnaire-based assessment approaches tend to introduce a high level of subjectivity, while lacking the comprehensive view of the person's life and status that only continuous monitoring can provide. This paper presents a framework that provides objective observations regarding the completion of daily tasks of the person with dementia, focusing on the differences between mild cognitive impairment (MCI) and Alzheimer disease (AD) patients. More specifically, a sensor-based framework is proposed as a remedy to support clinicians' assessments and eventually patients by semantically integrating and analyzing multi-sensory data. However, approaches with more than one sensors lead to three main problems: (i) how we can analyze the raw data from the sensors, (ii) how clinicians can use all the sensors efficiently during interventions and (iii) how results could be visualized in order to be valuable for the clinicians' assessments. The proposed system tries to address the above problems by employing knowledge-driven interpretation techniques based on Semantic Web technologies to analyze the sensors' recorded data, visualizing it through a simple clinician interface. In order to evaluate the system, a pilot study was conducted with 59 patients. The main goal of this pilot study is to identify whether the system is able to identify differences between MCI and AD patients and eventually verify the results of the clinical assessment. In the following sections, we present the theoretical background of the study (Sect. 2), the system architecture (Sect. 3) and finally the pilot study (Sect. 4).

2 Related Work

Lately, there has been a mounting interest worldwide to deploy pervasive computing technologies for advancing well-being and healthcare through remote monitoring and management services at the point of need. Within this line of research for remote healthcare, special attention has been given to the promotion of home-based continuous support and care of people with chronic conditions, to sustain their independence and eliminate the need for early hospitalisation by ensuring continuous monitoring for timely intervention. Dementia, being one of the leading causes of chronic poor health and disability for the aging population, has been the subject of a number of such research works. Remote management of people with dementia involves providing tools and services that allow them to be independent. One category of such tools includes external memory aids [1] that may be calendars, diaries, alarm watches, whiteboards, notebooks, and timers. It has been shown that such devices help people with mild dementia maintain an account of their daily life [2]. Another common practice is life story work, where reminiscing activities in one's present life provides a sense of control over the past, present and future [3]. Home environment [4], which aims to promote the use of embedded technology to support a person within their own living environment and extend the period of time they can remain in their own home prior to institutionalization. Within such an environment, it is common to find sensors. Sensors are the devices, which can record information about the person or the environment.

Chan et al. [5] monitor activities and features such as nighttime activities between 9 p.m. and 7 a.m., including in-bed restlessness, visiting toilets and going out. The authors find a correlation between in-bed restlessness and getting up or going out of the bedroom. Suzuki et al. [6] used door/window and photo-electric sensors to detect when someone came within 2 m and a wattmeter to detect when the TV is on. A baseline rhythm was established using daily questionnaires and sensor output statistics for 12 days in order to determine a 'typical' day. Tyrer et al. [7] focused on activities and features like presence, restlessness (using time and event-frequency). The restlessness data were related to the activities recorded by the residents (e.g., increase in bedroom activity after a knee surgery or frequent bathroom visits after change of medication).

Most of the efforts described above tend to concentrate on specific aspects, such as health status monitoring, alerts and reminders based on scheduled activities (e.g. medicine taking, training activities, etc.), and not on the overall behavior and daily activities modeling. This paper presents a system, in which multiple sensor outputs and their interpretation, through new reasoning mechanisms and knowledge structures, present clinicians with a comprehensive outlook of the person's ability in daily activities. The system is used in a lab environment during a specific clinical intervention (protocol), which is presented in the following section.

3 The Proposed Framework

A multi-sensor framework should provide not only efficient user interfaces, but also a flexible and robust underlying infrastructure to support data and functions presented to the user. In the field of Ambient Intelligence (AmI), and especially Ambient Assisted Living (AAL), this requirement for flexibility on the application layer is resolved by the Service-Oriented Architecture (SOA). For the clinical intervention, we designed and developed such a service-oriented system, based on the Web Service technology. Overall, the system architecture involves three layers: the hardware, middleware and application layer. The hardware layer includes all sensors and devices. The core of the system provides the necessary abstractions and a universal interface for the application layer to access functions. To do so, the middleware incorporates core modules that deal with each device independently. Consequently, it provides a universal homogeneous interface based on the WSDL[1] standard. The framework supports both pull-based and push-based real-time sensors for energy and object movement detection. **Sensors.** The sensors currently included in the framework support a variety of data formats and are non-intrusive, whether ambient or wearable. In this scenario, an *ambient depth camera*[2] is placed to survey the whole room, collecting both image and depth data. The *Plug* sensors[3] are attached to electronic devices, i.e. to the tea kettle and radio in this intervention, to collect power consumption data. *Tags*[4] are attached to objects of interest, i.e. tea cup, kettle, drug-box, watering can, folder of bills and phone, capturing motion events and consequently, providing information about the objects used during the protocol. Regarding wearable sensors, the intervention employs the *wristwatch*[5] that measures moving intensity during directed and semi-directed activities, and a *wearable wireless microphone* to record the participant's voice during directed activities and vocal tasks. **Data Retrieval and Storage.** One of the main technical obstacles in concurrent AAL systems, is to unify and timely coordinate sensor data retrieval, synchronicity and homogeneity. Indeed, the sensors used in our intervention present various data forms and especially both real-time and asynchronous data transfer. A dedicated module interfaces with each sensor, complying with its own API and platform dependencies. The ambient camera module continuously stores image and depth frames from the device's live feed. In addition, the module developed for Tags is push-based, sending certain sensor events to the system for storage, such as motion detection. The plugs on the other hand, are pull-based. Therefore, the module for plugs constantly polls them for power consumption data. The module for the microphone is also able to start and retrieve a recording on-demand. While, the aforementioned sensors are always online to the system, the wristwatch has to be connected to start the data transfer. However, this process is also aided by the system, as the wristwatch module is able to automatically retrieve and process wristwatch data upon connection.

Analysis. After data has been either streamed online to the system (i.e. camera and audio recordings, plug consumption data and motion detection events) or transferred offline (i.e. wristwatch accelerometer values), they must be further analyzed. The purpose of low-level sensor data analysis is to extract higher-level, more meaningful observations, by means of aggregation and machine learning techniques.

In detail, camera depth data is used by *Complex Activity Recognition (CAR)* [8] algorithms to provide location-based events e.g. the participant is in the tea zone. It also detects successful directed activities and logs their duration (i.e. walking). Image data from the same camera is used by another set of techniques, namely *Human Activity Recognition (HAR)* [9], which perform learning to detect high-level activities such as preparing tea. The energy consumption data stream is processed in

[1]WSDL: http://www.w3.org/TR/wsdl.

[2]Xtion Pro: http://www.asus.com/Multimedia/Xtion_PRO/.

[3]Plugwise sensors: https://www.plugwise.nl/.

[4]Wireless Sensor Tag System: https://www.plugwise.nl/.

[5]DTI-2, provided by Philips, www.philips.com.

real-time and based on certain per-device thresholds, pushes detected activities i.e. `KettleOn` and `RadioOn`. Similarly, analyzing the stream of tag sensor events results in the detection of object movement events, such as `KettleMoved`, `CupMoved` etc. Audio data is processed by Offline Speech Analysis (OSA) algorithms [10], to provide clustering for the patients as either *healthy*, *MCI* or *Alzheimer's Disease*. Raw accelerometer readings from the wristwatch are aggregated into a single numerical measurement, which signifies the participant's per minute *moving intensity*. The analysis layer encapsulates two additional, software-based modules: Knowledge Base Manager (KBM) and Semantic Interpretation (SI). The former module is able to parse detected activities and measurements under a common exchange model and map them into RDF format for storage in the Knowledge Base (KB). This process allows for the SI module to reason upon and fuse existing observations in order to extract higher-level activities. All analysis modules are exposed through universal WSDL endpoints. A controller module, implemented as the application backend, is responsible to timely invoke analysis and functions and guide data flow for storage in the KB. For example, the controller invokes microphone recordings for audio analysis, then initiates the OSA module, stores results through KBM and invokes SI when the lab session is complete.

Knowledge-driven Semantic Interpretation. Semantic Web technologies and in particular RDF/OWL[6] ontologies, have been gaining increasing attention as a means for modelling and reasoning over contextual information and human activities in particular [11]. Formally founded in Description Logics [12], their expressiveness and level of formality make them well-suited for the open nature of context-aware computing [13–15]. For example, in OWL one can effectively model and reason over taxonomic knowledge. This is a desirable feature in pervasive applications where the need to model information at different levels of granularity and abstraction, so as to drive the derivation of successively further detailed contexts is particularly evident. In addition to the native reasoning services, ontologies are usually combined with rules [16], allowing to express richer semantic relations, e.g. temporal relations. Under this context, low-level information acquired from detectors, such as video cameras and contact sensors, is mapped to ontology class and property assertions, while high-level interpretations are inferred through the combination of ontology semantics and rules. The knowledge-driven interpretation services in our system consist of a hybrid reasoning architecture that combines the OWL reasoning paradigm and the execution of SPARQL[7] queries. More specifically, the native semantics of OWL is used to formally represent and integrate activity-related information originated from different data sources, whereas SPARQL queries further aggregate activities, describing the contextual conditions and the temporal relations that drive the derivation of complex activities, e.g. the complex activities of the protocol. In addition, SPARQL assessment queries validate the underlying activity models, detecting abnormal behaviors, such as missed

[6]http://www.w3.org/TR/owl2-overview/.

[7]http://www.w3.org/TR/sparql11-overview/.

protocol activities or activities with long duration, assisting clinicians in assessment. In the following sections, we describe the basic notions that underpin the activity modelling and interpretation capabilities of our framework.

Semantic Activity Models. Information regarding the low-level events of the domain (referred to as *observations*), such as motion sensor activations, detected zones as well as the complex (high-level) activities of the protocol, such as `WaterPlan` and `PrepareDrugBox`, are captured as class instances of a lightweight ontology model. Each event instance of the model can be linked to ranges of time through the `atTime` property, allowing the definition of the time interval the event took place. Moreover, the actor of the event, i.e. the person who triggers the sensor or performs the complex activity, is defined through the `involvedAgent` property.

Rule-based Interpretation and Assessment. Given a set of observations e.g. a person's location, posture or motion sensor activations, the aim of the underlying reasoning framework is to meaningfully aggregate, correlate and interpret the information in order to elicit an understanding of the situation and to infer clinically relevant activities and problems. This is achieved through a set of domain-dependent SPARQL queries that encapsulate knowledge regarding: (a) The contextual dependencies among the events that define the complex activities of the domain, and (b) Clinical knowledge about abnormal behaviors that need to be highlighted to the clinicians after each protocol. For example, the recognition of the `PrepareTea` activity requires the activation of the kettle and cup motion sensors, the activation of the plug sensor of the kettle and the visual detection of the person inside the tea zone.

In addition, clinical knowledge about potentially abnormal patient behaviors pertinent to the protocol activities is defined in terms of assessment SPARQL queries. For example, activities with longer duration than usual are detected by rules and highlighted to the clinicians, such as a `ReadArticle` activity with duration longer than 52 s. These thresholds were suggested by the clinicians and corresponds to the average time needed to complete the tasks.

The Clinician Interface. One of the main limitations of the systems presented in the related work section is that they require the presence of a technician during a protocol. Considering this issue, we designed and developed the clinician interface based on the following main principles and goals: (i) The system should be easily operated by a novice computer user, (ii) The clinicians should be always informed about the current status of the protocol and the relevant instructions, (iii) The clinicians should easily operate the sensors and they should be informed about potential problems with the sensors, (iv) The clinicians should be informed about the results of each patient in a comprehensive way after the completion of the protocol. In order for the human computer interaction to be efficient, the tasks, procedures and methods that the clinician may perform with the system need to be structured in a logical and consistent manner. This means that the interface should be intuitive and the system should address the clinician's goal and objectives. Moreover, the number of actions that a user has to perform in accomplishing a task (even for technologically demanding tasks, like the operations of the sensors) should be minimized.

4 Methods

The participants. 59 participants (29 AD and 30 MCI) were recruited at the Day Care Alzheimer Center in Greece. Each participant gave informed consent before the assessment. The procedure has the approval of the Ethical Committee of the Greek Association of Alzheimer's Disease and Related Disorders.[8] Participants aged 60 or older. **The intervention.** The intervention was conducted in a lab environment in the Alzheimer day care center. Each participant starts with a regular consultation with a general practitioner, and then undergoes the ecological assessment, which is followed by a neuropsychological assessment. During the consultations, demographical and medical characteristics are gathered by means of widely used and generally recognized assessment tools. Using an instruction sheet, participants had to complete 8 daily-living-like activities in 15 min: *Make a phone call, Water a plant, Prepare tea, Listen to the radio, Read a magazine and write some answers, Establish account balance, Check the pill box, Exit the room.* The intervention consists of monitoring persons with dementia using the system's technology in order to provide a brief overview of their health status during consultation (cognition, behaviors and function), and to correlate the system data with the data collected using typical dementia care assessment tools. **Measures.** In order to examine the association of the monitored data and the clinical assessment, a comparison between the two groups (AD and MCI participants) was conducted. More specifically, T-test control was applied to patients' monitored successful attempts during the interventions. **Results.** In the following Table 1 the results of the t-test analysis for the successful attempts for the two groups is presented.

Table 1 Monitored data analysis

	Activities (successful attempts)	AD n = 29	MCI n = 30	t-test
1	Prepare tea	M = .17 SD = .38	M = .43 SD = .5	**p = 0.3**
2	Make a phone call	M = .41 SD = .57	M = .97 SD = .67	**p = 0.0**
3	Establish account balance	M = .28 SD = .65	M = .93 SD = .69	**p = 0.0**
4	Read article	M = .62 SD = .94	M = .93 SD = .74	p = .16
5	Listen to the radio	M = .034 SD = .19	M = .27 SD = .52	**p = .03**
6	Water the plant	M = .48 SD = .51	M = .87 SD = .35	**p = 0.01**
7	Leave the room	M = 1.1 SD = .84	M = 1.3 SD = .83	p = .37
8	Check the pill box	M = .35 SD = .67	M = .6 SD = .56	p = 0.12

[8]Greek Association of Alzheimer Disease and Relative Disorders (GAADRD). http://www.alzheimer-hellas.gr/english.php.

5 Conclusions

The presented study demonstrates a sensor-based system that is able to implement a specific clinical intervention to support clinicians in everyday practice for the assessment of dementia patients. The main goals of this paper are (a) to present the main technology components of the system and (b) to present the results of a pilot study regarding the patients' successful daily activities in a clinical protocol. The results of the pilot study reveals that in general the MCI patients carried out more activities than the AD patients. However, in three tasks there is no significant difference between the two groups. This can be explained because these three activities (e.g. check the pill box) were the easiest ones and almost every participant was able to accomplish. Consequently, in order to have a supportive and useful intervention the activities has to be carefully selected. Acceptability by clinicians depends on whether the monitoring system improves the effectiveness of their assessments and whether it reduces their work-load. The presented system supports their decision process without increasing their work load, summarizing clinically useful/relevant information rather than visualizing a vast amount of clinically irrelevant data.

Acknowledgment This work has been supported by the EU FP7 project Dem@Care: Dementia Ambient Care – Multi-Sensing Monitoring for Intelligent Remote Management and Decision Support under contract No. 288199.

References

1. M.L. Lee, A.K. Dey, Capturing and Reviewing Context in Memory Aids. Human-Computer Interaction Institute, Carnegie Mellon University, Pittsburgh, PA. Presented in: Workshop on Designing Technology for People with Cognitive Impairments (2006)
2. N. Kapur, E.L. Glisky, B.A. Wilson, External memory aids and computers in memory rehabilitation, in *The Hand-book of Memory Disorders* (Chichester, Wiley, 2002), pp. 757–784
3. S. Housden, Reminiscence and lifelong learning. Int. J. Comput. Healthcare 161–176 (2007)
4. J.C. Augusto, C.D. Nugentn, *Designing Smart Homes: The Role of Artificial Intelligence, LNAI 4008*. (Springer, Berlin, 2006)
5. M. Chan, E. Campo, D. Este`ve, Assessment of activity of elderly people using a home monitoring system. Int. J. Rehabil. Res. **28**:69–76 (2005)
6. R. Suzuki, S. Otake, T. Izutsu, M. Yoshida, T. Iwaya, Rhythm of daily living and detection of atypical days for elderly people living alone as determined with a monitoring system. J. Telemed. Telecare **12**, 208–214 (2006)
7. H.W. Tyrer, M.A. Aud, G. Alexander, M. Skubic, M. Rantz, Early detection of health changes in older adults. Conf. Proc. IEEE Eng. Med. Biol. Soc. 4045–4048 (2007)
8. R. Romdhane, C.F. Crispim, F. Bremond, M. Thonnat (2013) Activity recognition and uncertain knowledge in video scenes, in *2013 10th IEEE International Conference on Advanced Video and Signal Based Surveillance (AVSS)*, pp. 377–382
9. K. Avgerinakis, A. Briassouli, I. Kompatsiaris Recognition of activities of daily living for smart home environments, in *2013 9th International Conference on Intelligent Environments (IE)*, pp. 173–180 (2013)

10. A. Satt, A. Sorin, O. Toledo-Ronen, O. Barkan, I. Kompatsiaris, A. Kokonozi, M. Tsolaki, Evaluation of speech-based protocol for detection of early-stage dementia, in *INTERSPEECH*, pp. 1692–1696 (2013)
11. T. Tiberghien, M. Mokhtari, H. Aloulou, J. Biswas, Semantic reasoning in context-aware assistive environments to support ageing with dementia, in *International Semantic Web Conference*, pp. 212–227 (2012)
12. F. Baader, The description logic handbook: theory, implementation, and applications. (Cambridge University press, Cambridge, 2003)
13. D. Riboni, C. Bettini, COSAR: hybrid reasoning for context-aware activity recognition. Pers. Ubiquit. Comput. **15**(3), 271–289 (2011)
14. L. Chen, C. Nugent, Ontology-based activity recognition in intelligent pervasive environments. Int. J. Web Inf. Syst. **5**(4), 410–430 (2009)
15. G. Okeyo, L. Chen, H. Wang, R. Sterritt, Dynamic sensor data segmentation for real-time knowledge-driven activity recognition. Pervasive Mob. Comput. (2012)
16. G. Okeyo, L. Chen, H. Wang, R. Sterritt, A hybrid ontological and temporal approach for composite activity modelling, in *Trust, Security and Privacy in Computing and Communications*, pp. 763–1770 (2012)

Unified Fingerprinting/Ranging Localization for e-Healthcare Systems

Javier Prieto, Juan F. De Paz, Gabriel Villarrubia, Javier Bajo and Juan M. Corchado

Abstract Indoor localization constitutes one of the main pillars for the provision of context-aware services in e-Healthcare systems. Fingerprinting and ranging have traditionally been placed facing each other to meet the localization requirements. However, accurate fingerprinting may worth the exhaustive calibration effort in some critical areas while easy-to-deploy ranging can provide adequate accuracy for certain non-critical spaces. In this paper, we propose a framework and algorithm for seamless integration of both systems from the Bayesian perspective. We assessed the proposed framework with conventional WiFi devices in comparison to conventional implementations. The presented techniques exhibit a remarkable accuracy improvement while they avoid computationally exhaustive algorithms that impede real-time operation.

Keywords Bayesian data fusion · Fingerprinting · Ranging · RSS

1 Introduction

The development of a better condition of life for older adults has become a major task over the increasingly aged occidental population [10]. Information and communication technologies are acquiring a major role in this task since they can foster greater quality of life, autonomy and participation in social life of elderly people [4]. Such technologies have resulted in the irruption of a vast range of elderly care [4], home care [5] and e-Healthcare systems [2]. These systems are highly correlated with context, and hence, the knowledge of the location of elderly people can remarkably benefit the provision of these context-aware services [12]. Current posi-

J. Prieto (✉) · J.F. De Paz · G. Villarrubia · J.M. Corchado
BISITE Research Group, University of Salamanca, Edificio I+D+i C/ Espejo,
37007 Salamanca, Spain
e-mail: javierp@usal.es

J. Bajo
Departamento de Inteligencia Artificial, Facultad de Informática,
Universidad Politécnica de Madrid, Madrid, Spain

A. Mohamed et al. (eds.), *Ambient Intelligence - Software and Applications*,
Advances in Intelligent Systems and Computing 376,
DOI 10.1007/978-3-319-19695-4_23

tioning techniques that rely on global navigation satellite systems (GNSS) provide suitable performances in open areas [6]. However, there is no alternative technique with analogous performance and affordable complexity in harsh environments [11]. The proposed alternatives can be coarsely classified into fingerprinting and ranging localization techniques [12].

Fingerprinting techniques determine the position of a mobile target from location-dependent information provided by offline and online measurements [12]. In the offline phase, different features from the transmitted signals in the wireless network are stored at several locations to form a database of location fingerprints. Then, the position is estimated by comparison of the database with the values received by the target in the online phase (i.e., with its fingerprint). Fingerprinting techniques involve two major drawbacks: they require an arduous calibration offline phase and are very sensitive to fast environmental changes [12].

Ranging techniques determine the position of a mobile target from range-related information provided by received signal-strength (RSS) or time-of-arrival (TOA) measurements [11]. In a first stage, the distance to a set of anchors with known positions is estimated from the signals transmitted to the target. Then, the position is estimated by means of a process known as trilateration (i.e., intersection of circles). Ranging techniques suffer from two dominant limitations: their accuracy is far from fingerprinting methods and falls down under the multipath and non-line-of-sight (NLOS) conditions of harsh environments [6].

Strengths and weaknesses of fingerprinting and ranging localization have inevitably focused the challenge in developing hybrid systems without substantially increasing complexity and cost. Such solutions will enable fine localization via fingerprinting methods in places where the accuracy is critical or the database can be frequently updated, and coarse localization via ranging methods in areas where there is no database or it has become obsolete. In [8] fingerprint- and TOA-based methods are coupled to localize UWB devices from a maximum-likelihood (ML) perspective; in [9] fingerprint- and RSS-based techniques are fused by using RFID tags/readers and the computationally expensive particle filter; whereas in [3] fingerprint-based localization and channel-estimation tracking are combined to localize UWB devices via extended Kalman filter (EKF).

In this paper, we propose a framework for unified fingerprinting/ranging based on Bayesian data fusion. Such framework integrate position-related measurements from the first and range-related measurements from the second. Moreover, it considers the dynamic nature of the target's position and accommodates any other position-related information. We further derive algorithms to implement such framework based on the unscented Kalman filter (UKF) that allows for efficient computation over a Smartphone, facilitating its integration under previously proposed e-Healthcare solutions [12].

The rest of the paper is organized as follows: Sect. 2 exposes the system specification for its integration within an e-Healthcare platform; Sect. 3 presents the framework for unified data fusion of fingerprinting/ranging measurements; Sect. 4 assesses the proposed scheme by an experimental case study; and finally, Sect. 5 summarizes the conclusions drawn from the research.

2 System Specification

This section provides a general overview of the requirements of the localization framework implemented within a previously deployed e-Healthcare system [12].

The localization algorithm will be integrated within a monitoring and tracking platform for people with medical problems [12]. The mentioned platform is based on a virtual organization of agents that monitors user's information. The virtual organization was created with the PANGEA platform that facilitates the development of agents in light devices and the integration of different hardware [14]. Within this platform, the localization role is played in the home care organization. However, the localization algorithm can be executed in the elderly user's device (i.e., in a Smartphone) or in a centralized server (i.e., in the cloud). In the former case, the complexity limitations are imposed by the Smartphone's memory and CPU. In the latter case, the complexity constraints are determined by the number of simultaneous users and the response time. Consequently, we will avoid highly complexity demanding fusion algorithms such as particle filters or Gaussian mixture filters [9, 11].

The monitoring and tracking platform utilizes WiFi as the underlying technology to provide indoor localization. WiFi technology is more accessible and less expensive than other alternative technologies such as RFID or UWB, and has a longer range and larger bandwidth than ZigBee or Bluetooth. From signals transmitted in the WiFi network, we can easily extract the RSS metric, while time- or angle-related measurements imply additional complexities and costs [11]. Hence, we will employ the RSS metric for fingerprinting and ranging.

Figure 1 depicts the flowchart for the whole localization process. In the offline phase, we store the fingerprints for the most critical areas where we desire a more accurate localization. In the online phase, we estimate the position by means of Bayesian data fusion where fingerprinting/ranging models are chosen if the prediction is inside/outside the stored area, respectively. Note that this scheme accommodates other position-related information given by diverse devices such as the Smartphone's GPS receiver or a foot-mounted inertial measurement unit (IMU).

3 Bayesian Data Fusion

In this section, we formulate the problem of estimating the position of a mobile agent in a two-dimensional scenario by fusing information from different position-related measurements. In order to do that, we collect measurements, $\{\mathbf{y}_k\}_{k\in\mathbb{N}}$, at discrete time instants, $t_{k\in\mathbb{N}}$. From these measurements, we estimate the state vector, $\{\mathbf{x}_k\}_{k\in\mathbb{N}}$. In addition to the information conveyed by the measurements, the fact that the sequence of positions is highly correlated in time can also be used as another source of information. Next, we determine the entries to state and measurement vectors and define the models for the fusion of time-evolution and measuring information.

The state vector contains the position and its first derivatives so that $\{\mathbf{x}_k\}_{k\in\mathbb{N}}$ is a Markov chain (i.e., the current state only depends on the previous one) [1, 11]. In this paper, $\mathbf{x}_k = [\mathbf{p}_k, \mathbf{v}_k, \mathbf{a}_k] \in \mathbb{R}^6$, where $\mathbf{p}_k \in \mathbb{R}^2$ is the target's position, $\mathbf{v}_k \in \mathbb{R}^2$ its velocity, and $\mathbf{a}_k \in \mathbb{R}^2$ its acceleration.[1]

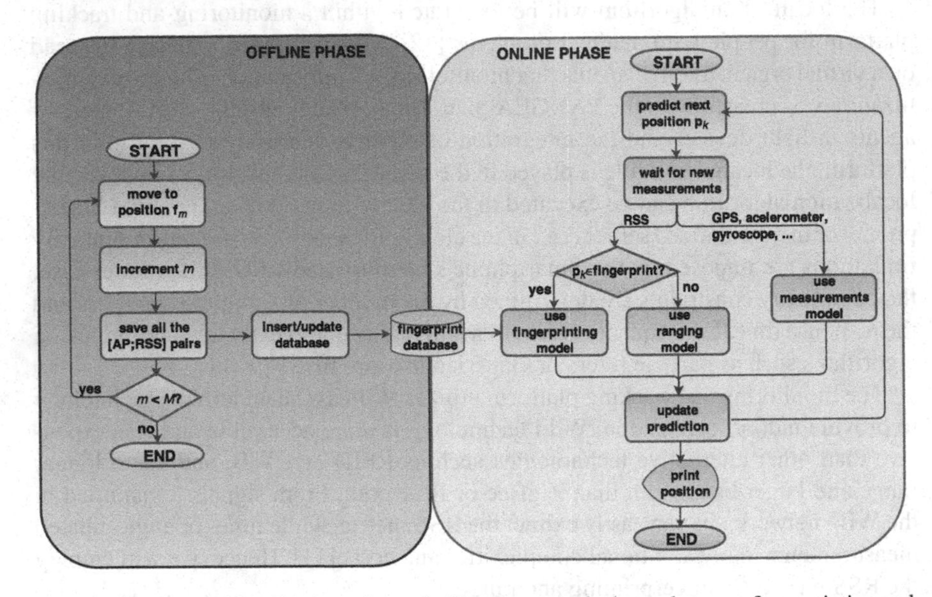

Fig. 1 The proposed localization scheme facilitates a smooth change between fingerprinting and ranging algorithms and the integration of GPS or inertial measurements

The measurement vector conveys any position-related information received at time instant t_k (i.e., its dimension may be different to the previous one). In this paper, $\mathbf{y}_k = \mathbf{y}_k^f \in \mathbb{R}^{L_k}$, or $\mathbf{y}_k = \mathbf{y}_k^s \in \mathbb{R}^{L_k}$, depending on whether we receive the RSS measurements within a critical area (i.e., fingerprinting) or not (i.e., ranging), respectively, where $L_k \in \mathbb{R}$ is the number of anchors visible at that particular moment.[2]

With the defined measurements and state vectors, it can be assumed that given the current state vector, $\mathbf{x}_k$, the measurement vector, $\mathbf{y}_k$, is independent of all previous and future states and measurements [1]. Therefore, we can build a hidden Markov model that leads to two kinds of dependence between the random variables: the relationship between the state vector in time t_k and the state vector in time t_{k-1}, i.e., $p(\mathbf{x}_k|\mathbf{x}_{k-1})$, called *dynamic model*; and the relationship between the measurements and the state vector in each time, i.e., $p(\mathbf{y}_k|\mathbf{x}_k)$, called *measurements model* [1, 11]. We define both models in the next subsections.

[1]In dead-reckoning systems, it is common to employ a foot-mounted IMU. In this case, the state vector has to be augmented to include the IMU's orientation, its derivatives and measurement biases [13].

[2]In dead-reckoning systems, it is common to gather specific force and angular velocity measurements. In this case, the measurement vector has to be augmented to include their respective values [13].

3.1 Dynamic Model

In the following, we define the model that conveys the information provided by the evolution in time of the state vector.

Given the position, velocity and acceleration at time t_{k-1}, $\mathbf{p}_{k-1}$, $\mathbf{v}_{k-1}$ and $\mathbf{a}_{k-1}$, we can approximate their values in time t_k, $\mathbf{p}_k$, $\mathbf{v}_k$ and $\mathbf{a}_k$, by means of their Taylor series expansion as [1],

$$\begin{bmatrix} \mathbf{p}_k \\ \mathbf{v}_k \\ \mathbf{a}_k \end{bmatrix} = \begin{pmatrix} \mathbf{I}_2 & \Delta_k \mathbf{I}_2 & \frac{\Delta_k^2}{2}\mathbf{I}_2 \\ \mathbf{0} & \mathbf{I}_2 & \Delta_k \mathbf{I}_2 \\ \mathbf{0} & \mathbf{0} & \mathbf{I}_2 \end{pmatrix} \begin{bmatrix} \mathbf{p}_{k-1} \\ \mathbf{v}_{k-1} \\ \mathbf{a}_{k-1} \end{bmatrix} + \mathbf{n}_k^d = \mathbf{F}_k \begin{bmatrix} \mathbf{p}_{k-1} \\ \mathbf{v}_{k-1} \\ \mathbf{a}_{k-1} \end{bmatrix} + \mathbf{n}_k^d, \quad (1)$$

where $\mathbf{I}_n \in \mathbb{R}^{n \times n}$ denotes the $n \times n$ identity matrix, $\Delta_k = (t_k - t_{k-1}) \in \mathbb{R}$ is the sampling interval, $\mathbf{F}_k \in \mathbb{R}^{6 \times 6}$ is the transition matrix, and $\mathbf{n}_k^d \in \mathbb{R}^6$ is the error term where the most common is to model it as white Gaussian noise (i.e., a discrete Wiener process). Therefore, the dynamic model is given by,

$$p(\mathbf{x}_k|\mathbf{x}_{k-1}) = \varphi(\mathbf{x}_k; \mathbf{F}_k\mathbf{x}_{k-1}, \mathbf{\Sigma}_k^d) \quad (2)$$

where $\varphi(\mathbf{x}; \boldsymbol{\mu}, \mathbf{\Sigma})$ denotes the probability density function of a random vector $\mathbf{x} \sim \mathcal{N}(\boldsymbol{\mu}, \mathbf{\Sigma})$ and $\mathbf{\Sigma}_k^d \in \mathbb{R}^{6 \times 6}$ is the covariance matrix corresponding to $\mathbf{n}_k^d$.

3.2 Measurements Model

In the following, we describe the models for the relationship between position- or range-related measurements and the state vector.

RSS-Fingerprinting Measurements. We assume that the RSS corresponding to the region of the map associated to fingerprint $\mathbf{f}_m \in \mathbb{R}^2$ follows a Gaussian distribution. Therefore, the likelihood function of such fingerprint is given by,

$$p(\mathbf{y}_k^f|\mathbf{f}_m) = \prod_{l=1}^{L_k} \varphi\left(y_k^{(l)}; \overline{\mathbf{y}}_m^{(l)}, \sigma_m^{(l)}/\sqrt{S_m^{(l)}}\right) \quad (3)$$

where $y_k^{(l)}$ is an RSS measurement received form the lth anchor in t_k, $\overline{\mathbf{y}}_m^{(l)}$ is the sample mean of the $S_m^{(l)}$ RSS measurements from such anchor stored in the mth fingerprint, and $\sigma_m^{(l)}$ is their sample standard deviation. Therefore, we can approximate the measurements model by a mixture of the individual likelihoods at every point of the set $\{\mathbf{f}_m\}_{m=1}^M$ as,

$$p(\mathbf{y}_k^f|\mathbf{p}_k) \approx \sum_{m=1}^{M} p(\mathbf{y}_k^f|\mathbf{f}_m)\varphi(\mathbf{p}_k;\mathbf{f}_m,h^2\mathbf{I}_2) \approx \sum_{m=1}^{M} \omega_k^{(m)}\varphi(\mathbf{p}_k;\mathbf{f}_m,h^2\mathbf{I}_2) \tag{4}$$

where we have selected a Gaussian kernel with bandwidth h to model each region of the map and approximate the likelihood by a continuous function.

RSS-Ranging Measurements. The RSS values are influenced, among other factors, by the distance between target and anchors. This attenuation is proportional to the inverse of the distance raised to a path-loss exponent [11]. In logarithmic units, we have that for the lth anchor with position $\mathbf{p}^{(l)} \in \mathbb{R}^2$,

$$y_k^s = \alpha - 10\beta \log_{10} \|\mathbf{p}^{(l)} - \mathbf{p}_k\| + n_k^s \tag{5}$$

where $\alpha \in \mathbb{R}$ is a constant that depends on several factors such as fast and slow fading, gains in transmitter and receiver antennas and the transmitted power, and $\beta \in \mathbb{R}$ is the path-loss exponent that can be dynamically obtained or trained in each scenario. Finally, n_k^s is a Gaussian noise term caused by shadowing [11]. Therefore, the corresponding likelihood function is given by,

$$p(y_k^s|\mathbf{p}_k) = \varphi(y_k^s; \alpha - 10\beta \log_{10} \|\mathbf{p}^{(l)} - \mathbf{p}_k\|, \sigma_k^p) \tag{6}$$

where $\sigma_k^s \in \mathbb{R}$ is the standard deviation corresponding to n_k^s.

4 Performance Evaluation

The goal of this section is to quantify the performance of the localization framework described in Sect. 3 that may help to a better monitoring and tracking of elderly people. The system is evaluated in the experimental case study of a pedestrian walking with a Smartphone that collects RSS measurements from the WiFi network. In the following, we describe the set-up for the experiments and present the performance results.

4.1 Experimental Set-Up

To obtain the localization results we utilized dynamic and measurements models described in Sect. 3.[3] The complexity constraints mentioned in Sect. 2 led to the election of a Kalman-like solution [7]. The lack of linearity in the models implied the use of a suboptimal solution, where the most common is the EKF. For rang-

[3] We also added zero-mean Gaussian priors for velocity and acceleration.

ing localization, we selected the UKF since it better captures higher order moments caused by non-linearities and avoids computation of Jacobian and Hessian matrices [7]. For fingerprinting localization, we predicted via Kalman filter (KF) and updated by multiplying the prediction with the Guassian mixture likelihood given by (4) and approximating the result by a single Gaussian.

The mobile target was a pedestrian with a Smartphone that covered the path shown in Fig. 2b with several people walking around. The total length of the path was approximately 110 m, implying a total time of 2.5 min. For fingerprinting localization, the database was created by storing at least 10 RSS values from all the detectable access points (up to 25) in the fingerprints marked in Fig. 2a. For the ranging localization, we employed 16 RSS measurements per point from the 4 access points plotted in Fig. 2a. All the RSS measurements were considerably affected by NLOS and multipath propagation conditions.

We compare our results against a conventional implementation based on Bayesian networks for fingerprinting [2] and ML for ranging [11].

4.2 Results and Discussion

Figure 2b shows the localization results in the mentioned path. For the proposed approach, the root-mean-square-error (RMSE) was 0.79 m in fingerprinting, 4.04 m in ranging, and 2.55 m in total. For the conventional approach, the RMSE was 1.48 m in fingerprinting, 5.88 m in ranging, and 3.79 m in total.

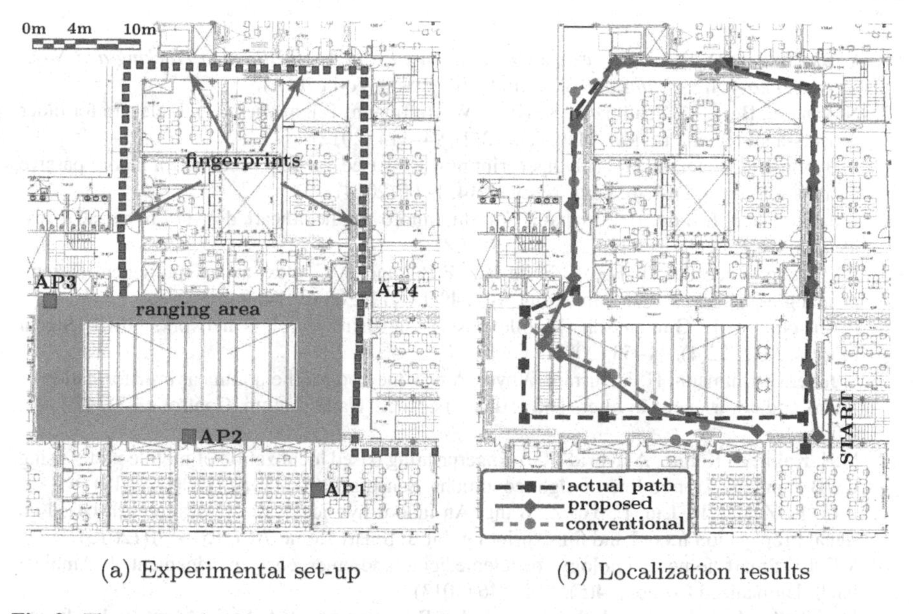

(a) Experimental set-up (b) Localization results

Fig. 2 The proposed unified framework provides accurate localization via fingerprinting for critical areas and ready-to-use localization via ranging for non-critical spaces

From Fig. 2 we can point out that: (1) the proposed framework facilitates the shift from accurate fingerprinting to coarse ranging; (2) the Bayesian approach noticeably improves ML in ranging localization; and (3) fingerprinting outperforms ranging in WiFi networks while requiring greater calibration effort.

5 Conclusion

This paper has presented a principled framework and efficient algorithm for unifying fingerprinting and ranging localization in e-Healthcare systems. We have defined the framework from the Bayesian perspective that allows for the inclusion of position-related information coming from heterogeneous sources. We have implemented the algorithm via UKF that holds promise for effective fingerprinting/ranging fusion without a substantial increment in complexity. Under NLOS and multipath conditions, the presented techniques obtained an error in position estimation of 2.5 m along a 110-meter-long path, remarkably outperforming conventional fingerprinting/ranging implementations.

Acknowledgments This work has been supported by the Spanish Government through the project iHAS (grant TIN2012-36586-C01/C02/C03) and FEDER funds.

References

1. Y. Bar-Shalom, X.R. Li, T. Kirubarajan, *Estimation with Applications to Tracking and Navigation: Theory Algorithms and Software* (Wiley, New York, 2001)
2. N. Van den Berg, M. Schumann, K. Kraft, W. Hoffmann, Telemedicine and telecare for older patients-a systematic review. Maturitas **73**(2), 94–114 (2012)
3. S. Bybordi, L. Reggiani, Hybrid fingerprinting-EKF based tracking schemes for indoor passive localization. Int. J. Distrib. Sens. Netw. **2014**, 1–11 (2014)
4. J.M. Corchado, J. Bajo, A. Abraham, GerAmi: improving healthcare delivery in geriatric residences. IEEE Intell. Syst. **23**(2), 19–25 (2008)
5. J.A. Fraile, Y. de Paz, J. Bajo, J.F. de Paz, B.P. Lancho, Context-aware multiagent system: planning home care tasks. Knowl. Inf. Syst. **40**(1), 171–203 (2014)
6. F. Gustafsson, F. Gunnarsson, Mobile positioning using wireless networks. IEEE Signal Process. Mag. **22**(4), 41–53 (2005)
7. S. Julier, J. Uhlmann, H.F. Durrant-Whyte, A new method for the nonlinear transformation of means and covariances in filters and estimators. IEEE Trans. Autom. Control **45**(3), 477–482 (2000)
8. M.H. Kabir, R. Kohno, A hybrid TOA-fingerprinting based localization of mobile nodes using UWB signaling for non line-of-sight conditions. Sensors **12**(8), 11187–11204 (2012)
9. J. Li, B. Zhang, H. Liu, L. Yu, Z. Wang, An indoor hybrid localization approach based on signal propagation model and fingerprinting. Int. J. Smart Home **7**(6), 157–170 (2013)
10. K.F. Li, Smart home technology for telemedicine and emergency management. J. Ambient Intell. Humanized Comput. **4**(5), 535–546 (2013)
11. J. Prieto, S. Mazuelas, A. Bahillo, P. Fernández, R.M. Lorenzo, E.J. Abril, Adaptive data fusion for wireless localization in harsh environments. IEEE Tran. Sig. Proc. **60**(4), 1585–1596 (2012)

12. G. Villarrubia, J.F. de Paz, J. Bajo, J.M. Corchado, Monitoring and detection platform to prevent anomalous situations in home care. Sensors **14**(6), 9900–9921 (2014)
13. F. Zampella, A. Bahillo, J. Prieto, A.R. Jiménez, F. Seco, Pedestrian navigation fusing inertial and RSS/TOF measurements with adaptive movement/measurement models: experimental evaluation and theoretical limits. Sens. Actuators A: Phys. **203**, 249–260 (2013)
14. C. Zato, G. Villarrubia, A. Sánchez, I. Barri, E. Rubión, A. Fernández, C. Rebate, J.A. Cabo, T. Álamos, J. Sanz, J. Seco, J. Bajo, J.M. Corchado, PANGEA—platform for automatic coNstruction of orGanizations of intElligent agents, in *Distributed Computing and Artificial Intelligence*. Advances in Intelligent and Soft Computing, vol. 151 (Springer, Berlin, 2012), pp. 229–239

My Kinect Is Looking at Me - Application to Rehabilitation

Miguel Oliver, Antonio Fernández-Caballero, Pascual González, José Pascual Molina and Francisco Montero

Abstract This paper studies the feasibility of using two Kinect sensors, compared to only one, as part of a system of computer-aided rehabilitation. The use of multiple sensors to collect and further process the movements of users, is initially an advantage over the use of a single sensor, but the interference due to the co-existence of two sensors in the same scene must be considered. The purpose of this paper is to determine how overlapping of several beams of infrared light affect the capture of users, in function of the angle of incidence and the distance to the targets. The paper also examines whether the use of two Kinect sensors increases the accuracy of the data collected and the range of action of the sensors.

Keywords Kinect sensors · Rehabilitation system · Capture precision

1 Introduction

Computer-based assistance and rehabilitation systems are becoming popular in recent years (e.g. [1–3]). These systems have a distinct advantage over traditional systems rehabilitation, where rehabilitation specialists have to constantly monitor the fulfillment of the patients' exercises. An article [4] shows that the accuracy of depth sensors as *Microsoft Kinect* is sufficient for rehabilitation exercises. Another article [5] concludes that the accuracy is high enough for the device to be used in the fields of ergonomics, medicine and biometric analysis, among others. At first, rehabilitation support systems were devised with one single sensor for capturing the patient's body. An example [6] helps in the rehabilitation of people with motor problems.

M. Oliver · A. Fernández-Caballero (✉) · P. González · J.P. Molina · F. Montero
Universidad de Castilla-La Mancha, Instituto de Investigación en Informática de Albacete, 02071 Albacete, Spain

A. Fernández-Caballero · P. González · J.P. Molina · F. Montero
Universidad de Castilla-La Mancha, Departamento de Sistemas Informáticos, 02071 Albacete, Spain

A. Mohamed et al. (eds.), *Ambient Intelligence - Software and Applications*, Advances in Intelligent Systems and Computing 376,
DOI 10.1007/978-3-319-19695-4_24

Another example [7] adds a Wiimote sensor for the return of haptic sensations. A system [8] promotes a recreational component to motivate the user to continue rehabilitation and thus avoid abandonment. In addition, we have systems in operation as *KineLabs*, *Reflexion*, *Toyra*, *TeKi* and *VirtualRehab*.

Besides single depth camera systems, there are alternatives which deploy multiple cameras to capture better the user's body. An example [9] shows a multimodal rehabilitation system with multiple *iotracker* sensors that capture the patient's body from multiple angles. All rehabilitation systems require that the user is perfectly located for interaction. Furthermore, one system (sensor, display device, processing device) is needed for each user, which prevents usefulness in rehabilitation rooms where several patients are performing physical rehabilitation at the same time. Therefore, solutions are emerging in which a recognition system is introduced with multiple users and multiple Kinect sensors [10]. However, the main problem in the use of several sensors is the interference that may arise between the infrared beams sent by each sensor to determine the depth of the scene. In [11] a study of the accuracy of a Kinect sensor and an Asus Xtion PRO LIVE sensor is made. It concludes that at distances of 2 m the error is less than 2.5 cm. Another study on Kinect sensor accuracy [12] concludes that the accuracy of such sensors is sufficient for distances of 1–3 m from the sensor to the target.

It is worth noting an article [13] in which a study on the accuracy of two Kinects depth sensors and six PlayStation Eye cameras is performed. The study indicates that the accuracy of the depth sensors, pointing to the same object and displaced 108° between them is always less than 10 cm. Another article [14] presents a study on erroneous pixels in the depth map when using one versus two Kinect sensors. In the case of using two Kinect sensors which recognize the same scene, the erroneous pixels become almost double that when the study is done with just one single Kinect. The result of the tests shows a large entropy when the two sensors share a surface which reflects their infrared beams.

Thus, the aim of this paper is to present a study on the accuracy in the measures of a couple of Kinect sensors when subjected to interference between themselves. The paper has to determine how overlapping of several beams of infrared light affect the capture of users, in function of the angle of incidence and the distance to the targets. The paper also examines whether the use of two Kinect sensors increases the accuracy of the data collected and the range of action of the sensors.

2 Description of One Versus Two Kinect Sensors Experiments

The experiments consist of measuring the position of a user, located in the survey area of Kinect depth sensors and determining how the number and position of the capture devices affects the accuracy in the users positioning of the data obtained. The depth sensors emit a pattern of infrared light that bounces off nearby objects, and determine the distance to them by identifying the deformation of the infrared pattern. The devices have a field of view of 57.5° in horizontal and 43.5° in vertical, with a maximum viewing distance of about 4 m and a minimum distance of 80 cm from the sensors.

The proposed experimentation incorporates a mannequin that serves as the user. The test mannequin solves the problem of involuntary movement that humans usually perform. The mannequin is positioned at previously established marks on the floor and the distance is measured from the sensor to the mannequin's waist. The data collected by different Kinect sensors, through Microsoft's skeleton tracking algorithm, are transferred to a global coordinate system. Figure 1 details the four experiments. In each, there are three different settings which indicate the position adopted by the Kinect sensors, as well as the grid representing the measuring points where the mannequin is located. For each of these settings the measuring points are: 4 for the first, 9 for the second and 16 for the third, being each of these points 1 m away from neighboring points. In the case that the measurement point is not found within at least one recognized area, the point will not be tested. The position of each sensor is set to test how the relative position between the sensors affects the mutually produced interference noise, and how the distance between the sensors affects the said noise.

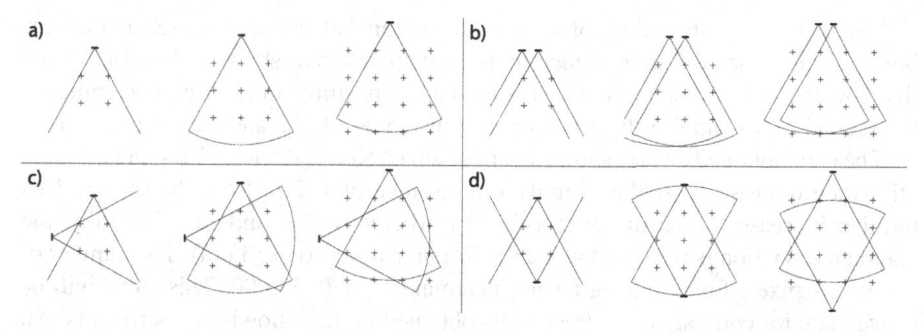

Fig. 1 Experiments performed. (a) Experiment with one single sensor. (b) Experiment with two sensors pointing to same direction. (c) Experiment with two perpendicular sensors. (d) Experiment with two opposing sensors

Precision is obtained by calculating erroneous pixels gotten from the sensors' depth map and the standard deviation of the location of the user. The number of erroneous pixels refers to positions of the depth image captured by the sensor, which distance cannot be obtained from the sensor. In each of the measuring points 100 samples of erroneous pixels of the scene and mannequin's location were collected. So, for each test a total of 2900 samples were taken. The experiments are performed in a space free of infrared light to increase measurement accuracy and avoid interference from other sources.

2.1 Experiment #1

This first experiment is performed with a single Kinect sensor pointing to front of the survey area, as shown in Fig. 1a. The mannequin, as in all other experiments, is oriented towards sensor. This experiment is performed to take a baseline for comparison with the rest of the experiments (with two Kinect sensors). A low (but not zero) number of erroneous measurements is expected. Although there is no interference between beams of different Kinect sensors, there are always erroneous measurements due to surfaces that do not properly reflect the infrared beam. The disparity between the view of the camera sensor and the view from the infrared emitter also involves errors. Furthermore, correct positioning of the mannequin is expected with a relatively small standard deviation. The user's position is determined by the next formula, where $x_{sensor1}$ and $y_{sensor1}$ are the positions, in X axis and Y axis, of the user which collects the first sensor, $x_{sensor2}2$ and $y_{sensor2}$ are the positions, in X axis and Y axis, of the user which collects the second sensor, dif_h is the horizontal separation between sensors, and dif_v is the vertical separation between sensors:

$$x_{user} = x_{sensor1} \quad ; \quad y_{user} = y_{sensor1}$$

The table shows the values of standard deviation and erroneous pixels in measuring positions. The data are provided in the same position as shown in Fig. 1a. So, the first row represents distances of 1 m from the test position and sensor, and similarly the second, third and fourth rows represent distances of 2, 3 and 4 m respectively.

The data obtained clearly show the operation of Kinect device. The standard deviations are quite small (below 1 mm) at a distance of 1–2 m from the sensor. This number increases when the distance is 3 m (between 1 m and 1 cm). Finally, the standard deviation is triggered when the distance is 4 m (over 1 cm). The number of erroneous pixels found during testing is around 50,000–53,000. This value will be a reference for comparison of the results obtained in the following experiments. All this information is shown in Table 1.

Table 1 Results of standard deviation in user location (in centimeters) and erroneous pixels – experiment #1

Conf.1		Conf.2			Conf.3			
0,10	0,02	-	0,03	-	-	0,02	0,05	-
0,03	0,06	0,02	0,03	0,13	-	0,03	0,03	-
		0,12	0,08	0,18	0,22	0,13	0,12	0,36
					1,14	4,71	1,31	3,22

Conf.1		Conf.2			Conf.3			
58	43	-	53	-	-	63	46	-
51	53	51	49	49	-	54	55	-
		49	48	50	53	53	53	53
					52	53	53	53

2.2 Experiment #2

Experiment #2 is performed with two Kinect sensors pointing in the same direction, as shown in Fig. 1b. This experiment should determine how two overlapping beams of infrared light, with the same recognition surface and orientation to the same direction affect their performance. Fairly high interference results are expected, since the sensors share the same area of reflection on the mannequin. Therefore, the internal Kinect pattern recognition algorithm is assumed to fail when identify patterns. The number of erroneous pixels and the standard deviation is expected to be higher than in experiment #1. The transformation of relative positions of each sensor to global positions is made by applying the following formula to each of data collected by the two sensors.:

$$x_{user} = \frac{x_{sensor1} + x_{sensor2} + dif_h}{2} \quad ; \quad y_{user} = \frac{y_{sensor1} + y_{sensor2}}{2}$$

It is found that the data worsen in almost all points of measurement. The use of two infrared emitters makes sensors interfere with each other. However, the results follow the same pattern as the first experiment results. With distances of 1 and 2 m the standard deviations are lower than 1 mm, with 3 m values distance between 1 mm and 1 cm are obtained, and when the distance is 4 m, the values are larger than 1 cm (see Table 2). The erroneous pixels in this experiment are also quite similar to those reported in the previous experiment. Nonetheless, when studying the data in more detail, we can see that erroneous pixels in this second experiment are, on average, slightly lower than those collected in the first. This is because the second Kinect senses the scene better and therefore the sensor recognizes more zones.

Table 2 Results of standard deviation in user location (in centimeters) and erroneous pixels – experiment #2

Conf.1		Conf.2			Conf.3			
0,07	0,03	0,03	0,06	-	-	0,09	0,03	-
0,07	0,07	0,10	0,05	0,22	0,05	0,06	0,07	-
		0,15	0,21	0,22	0,21	0,14	0,22	0,25
					3,44	3,79	5,51	4,69

Conf.1		Conf.2			Conf.3			
52	46	47	46	-	-	64	42	-
49	46	49	49	52	41	51	47	-
		46	47	46	49	49	49	55
					48	49	49	48

2.3 Experiment #3

This third experiment is also performed with two Kinect sensors; but now they are positioned perpendicularly to each other (yaw angle difference equal to 90°), as described in Fig. 1c. This experiment is thought to determine the effect of superposing two infrared beams perpendicularly reflecting on the same object. Smaller interference results are expected than during experiment #2. Although the number of sensors is the same in both cases, the surface on which the infrared pattern is reflected is not shared. Therefore, most of the pattern emitted by a sensor is not captured by the other. In this case, the second sensor has the coordinates inverted, relative to first sensor. This is represented in the following formula:

$$x_{user} = \frac{x_{sensor1} + y_{sensor2} + dif_h}{2} \quad ; \quad y_{user} = \frac{y_{sensor1} + x_{sensor2} + dif_v}{2}$$

When the two sensors are perpendicularly installed it can be seen how data greatly worsen. This is because two conditions coexist in this configuration (see Table 3):

- Firstly, an increase in average distance between the sensors and the mannequin is appreciated at certain points. In this configuration, the mannequin can be within a short distance of a sensor and a great distance from the other.
- The second circumstance, which is what most affects the data collected, is the poor precision of the Kinect sensors to recognize the mannequin's profile. The sensor positioned perpendicularly fails to capture the user completely, and many points of the mannequin have to be inferred, which reduces accuracy.

Regarding erroneous pixels it can be observed that the data obtained are almost always better than in the first experiment, and also improve compared to the second experiment. This is because the second sensor obtains an image with less erroneous pixels than the other sensor, as happened in the experiment #2. This is a clear indication that, while the sensor has a very clean image without a high amount of erroneous pixels, the location of the user depends on other factors such as how the user is positioned towards the sensor.

Table 3 Results of standard deviation in user location (in centimeters) and erroneous pixels – experiment #3

Conf.1		Conf.2			Conf.3			
0,04	0,14	-	0,06	0,23	-	0,02	0,26	4,72
0,31	0,04	0,02	0,08	0,12	0,55	0,05	0,32	2,04
		0,14	0,31	0,23	0,14	0,26	0,10	0,52
					1,02	6,25	2,01	2,05

Conf.1		Conf.2			Conf.3			
42	31	-	42	44	-	64	37	27
39	35	49	41	40	38	43	43	28
		49	39	39	45	43	42	41
					52	53	41	40

2.4 Experiment #4

Again, this experiment is performed with two Kinect sensors, but they are now fully confronted (yaw angle difference equal to 180°), as shown in Fig. 1d. This experiment is performed in order to determine how a user localization is affected when an infrared source from another sensor which emits its infrared light directly on the sensor. In this case, a very high interference result is expected since one of the sensors will project its infrared pattern on the lens of the other sensor, which disables the lecture of correct depth data. The user's position is now calculated as:

$$x_{user} = \frac{x_{sensor1} - x_{sensor2}}{2} \quad ; \quad y_{user} = \frac{y_{sensor1} + y_{sensor2} + dif_y}{2}$$

Data from the last test show indeed a worsening in almost all the studied positions. But as you can see, the worsening of the data is greater when the user is recognized from his/her back. Thus, it can be concluded that the sensor does a better job when the user is recognized frontally (see Table 4).

Table 4 Results of standard deviation in user location (in centimeters) and erroneous pixels – experiment #4

Conf.1		Conf.2			Conf.3			
0,10	0,04	0,27	0,10	0,66	0,75	0,36	0,24	1,74
0,13	0,08	0,08	0,02	0,45	0,32	0,10	0,06	1,72
		0,11	0,10	0,17	0,23	0,15	0,11	0,25
					2,31	2,92	0,83	3,14

Conf.1		Conf.2			Conf.3			
43	36	43	48	28	65	64	55	63
38	45	46	48	48	65	59	60	63
		49	53	50	55	60	61	53
					52	51	60	53

3 Discussion

In view of the experiments, it can be said that the number of erroneous pixels obtained is not a reliable indicator. While a very high number of erroneous pixels indicates that there may be a problem with the capture, a low number does not ensure that the capture is correct. Therefore, in real systems it is possible to make use of erroneous pixels as an indicator of possible errors in capturing the user, but you can not trust the accuracy of measurement in this factor only.

In general, the data provided indicate that the use of more than one Kinect sensor determines that the data collected from a user are worse than being taken up with one sensor placed optimally. But the problem in determining the optimal position of a sensor, when the contemplated interaction space is large and multiuser, is quite important. Either the interaction space of the user is restricted and the setup is indicated so that the user is always facing the sensor, or the use of several sensors is necessary. There is no advantage in using two sensors pointing in the same direction and covering the same interaction space, as described in the second experiment, in comparison to the placement of a single sensor. The user is viewed similarly by both sensors, so that the second sensor does not provide additional information. This configuration must be discarded in most real cases where it is needed to place multiple sensors.

The placement of a sensor perpendicular to another, as seen in the third experiment, can be a help in recognizing the user who interacts freely. Thus, there is always a sensor that captures the user completely, thus minimizing the occlusions caused by the user's own body. The information arising from this configuration can be treated subsequently, to reject the data obtained by the sensor that captures the user's profile, only keeping the other sensor's data. Finally, the placement of two opposite Kinect sensors (see the fourth experiment), may be useful when considering a free interaction area and it is necessary to capture the front to the user. As in the previous case, further processing is possible to determine which sensor is best suited to provide reliable user data.

4 Conclusions

The motivation for this article comes from the great interest in computerized rehabilitation systems. Increasingly, multi-camera systems are created that enhance the captured depth image of the patient. The main problem of these systems is interference arising between the different depth sensors. Therefore, this article has studied the use of two Kinect sensors in order to assist in rehabilitation therapies in the near future, and support research developed simultaneously (e.g. [7, 10]). The experiments have shown the operation of Kinect sensors. Also, the interference arising from the use of two sensors in the same interaction space, according to the angle of incidence of infrared light, and the relative position between sensors and the user's body, has been evaluated.

The use of more than one Kinect sensor, covering the same area of interaction, can be useful at times, as it has been mentioned. However, the placement of the sensors must be studied carefully in order to obtain an improvement in respect of using only one sensor. This article aims to be useful for decision making in placing sensors in depth-based multi-camera systems.

Acknowledgments This work was partially supported by Spanish Ministerio de Economía y Competitividad / FEDER under TIN2012-34003 and TIN2013-47074-C2-1-R grants, and through the FPU scholarship (FPU13/03141) from the Spanish Government.

References

1. I. Marcelino, D. Lopes, M. Reis, F. Silva, R. Laza, A. Pereira, Using the eServices platform for detecting behavior patterns deviation in the elderly assisted living: a case study. BioMed. Res. Int. **530828**, (2014)
2. E. Navarro, A. Fernández-Caballero, R. Martínez-Tomás, Intelligent multisensory systems in support of information society. Int. J. Syst. Sci. **45**, 711–713 (2014)
3. A. Fernández-Caballero, J.C. Castillo, M.T. López, J. Serrano-Cuerda, M.V. Sokolova, INT3-horus framework for multispectrum activity interpretation in intelligent environments. Expert Syst. Appl. **40**, 6715–6727 (2013)

4. A. Fernández, A. Susín, X. Lligadas, Biomechanical validation of upper-body and lower-body joint movements of kinect motion capture data for rehabilitation treatments, in *4th International Conference on Intelligent Networking and Collaborative Systems* 2012, pp. 656–661
5. B. Bonnechère, B. Jansen, P. Salvia, H. Bouzahouene, V. Sholukha, J. Cornelis, M. Rooze, S. Van Sint Jan, Determination of the precision and accuracy of morphological measurements using the KinectTM sensor: comparison with standard stereophotogrammetry, Ergonomics **57**, 622–631 (2014)
6. Y. Chang, S. Chen, J. Huang, A Kinect-based system for physical rehabilitation: a pilot study for young adults with motor disabilities. Res. Dev. Disabil. **32**, 2566–2570 (2011)
7. M. Oliver, J.P. Molina, F. Montero-Simarro, P. González, A. Fernández-Caballero, Wireless multisensory interaction in an intelligent rehabilitation environment. Ambient Intell. Softw. Appl. (2014), pp. 193–200
8. D. Freitas, A. Da Gama, L. Figueiredo, T. Chaves, D. Marques-Oliveira, V. Teichrieb, C. Araújo, Development and evaluation of a Kinect based motor rehabilitation game. Simposio Brasileiro de Jogos e Entretenimento Digital (2012), pp. 144–153
9. C. Schönauer, T. Pintaric, H. Kaufmann, S. Jansen-Kosterink, M. Vollenbroek-Hutten, Chronic pain rehabilitation with a serious game using multimodal input, in *The 2011 International Conference on Virtual Rehabilitation* 2011, pp. 1–8
10. M. Oliver, F. Montero-Simarro, A. Fernández-Caballero, P. González, J.P. Molina, RGB-D assistive technologies for acquired brain injury: description and assessment of user experience. Expert Syst. (2015). doi:10.1111/exsy.12096
11. H. Gonzalez-Jorge, B. Riveiro, E. Vazquez-Fernandez, J. Martínez-Sánchez, P. Arias, Metrological evaluation of microsoft Kinect and asus xtion sensors. Meas. J. Int. Meas. Confederation **46**, 1800–1806 (2013)
12. K. Khoshelham, S.O. Elberink, Accuracy and resolution of Kinect depth data for indoor mapping applications. Sensors **12**, 1437–1454 (2012)
13. D. Regazzoni, G. de Vecchi, C. Rizzi, RGB cams versus RGB-D sensors: low cost motion capture technologies performances and limitations. J. Manuf. Syst. **33**, 719–728 (2014)
14. T. Mallick, P.P. Das, A.K. Majumdar, Characterizations of noise in Kinect depth images: a review. IEEE Sens. J. **14**, 1731–1740 (2012)

Flying Depth Camera for Indoor Mapping and Localization

Lidia María Belmonte, José Carlos Castillo, Antonio Fernández-Caballero, Sergio Almansa-Valverde and R. Morales

Abstract This paper introduces a flying robot mapping and localization proposal from an onboard depth camera. The miniature flying robot is part of an ongoing project related to ambient assisted living and home health. The flying depth camera is used with a double function; firstly, as a range sensor for mapping from scratch during navigation, and secondly, as a gray-scale camera for localization. The Harris corner detection algorithm is implemented as key point detector for the creation and/or identification of indoor spatial relations. During the localization phase, the spatial relations created from detected corners in the mapping phase are compared to the corners identified in the map. The flying robot position is estimated by matching these spatial relations.

Keywords Ambient assisted living · Miniature flying robot · Flying depth camera · Mapping · Localization

1 Introduction

Autonomous navigation of flying robots in GPS-denied environments such as indoors requires that the flying robot be able to recognize the environment using external sensors [1]. Our research team is engaged in introducing miniature flying robots in ambient assisted living (e.g. [2–4]) and home health [5] applications. Now, dealing with the high amount of obstacles inherent to home facilities is a major challenge for flying vehicles [6]. This why mapping and localization at homes of the flying robot are extremely important challenges.

Mobile robot mapping techniques are usually classified according to the map representation and the underlying estimation technique [7, 8]. One popular map repre-

L.M. Belmonte (✉) · A. Fernández-Caballero · S. Almansa-Valverde · R. Morale
Escuela de Ingenieros Industriales, Universidad de Castilla-La Mancha,
02071 Albacete, Spain

J.C. Castillo
Robotics Lab, Universidad Carlos III de Madrid, 28911 Madrid, Spain

A. Mohamed et al. (eds.), *Ambient Intelligence - Software and Applications*,
Advances in Intelligent Systems and Computing 376,
DOI 10.1007/978-3-319-19695-4_25

sentation is the occupancy grid [9]. Such grid-based approaches are able to represent arbitrary objects [10]. However, such systems rely on predefined feature extractors. In this sense, scan matching approaches demonstrate to produce consistent maps [11]. Now, in mobile robot localization, an appearance-based approach for place recognition involves the matching of scenes based on selected features or landmarks observed within the current local map or sensor view. For each feature a descriptor vector that encodes the local area around that landmark is computed, thus allowing the comparison of features based on appearance. The combination of a location and descriptor vector is denominated a key point [12]. Localization then becomes a matter of identifying places and/or objects by associating key points, or deciding that a place/object was not previously seen.

Key points have been used in vision-based systems which use nearest neighbor voting [13] with SIFT [14] features. Another approach [15] uses a fast-Hessian detector to identify key points in an image and SURF [16]. In a review, detectors and local descriptors of local features for computer vision are described [17]. Another key point detector is the SUSAN corner detector [18] which finds corners based on the fraction of pixels that are similar to the center pixel in a circular region. Also, affine invariant feature extraction for localization in indoor environments, using the Harris corner detector [19] for local point detection, has been proposed [20].

Our proposal deals with specific characteristics of depth cameras [21] to solve the mapping and localization problem. Two constraints have been addressed. First, the generated map increases when new zones are discovered as the environment is unknown. And second, no landmark is introduced to facilitate the localization process since the robot flies in a real environment. Moreover, the looking forward depth camera is firstly used as a range sensor for mapping, and, afterwards as a gray-scale camera for localization. The depth camera, when used as a range sensor, provides a powerful tool for measuring the distance from the detected objects to the flying robot. On the other hand, the depth camera creates an image representing distances which enable extracting trackable features. These points in the space are the basis for the spatial relations used in the localization phase. Finally, the flying robot position is estimated in accordance to the localization information provided through the identified spatial relations. It should be noted that the proposal requires a flight at a constant height.

2 Mapping from Flying Robot

The depth camera is considered a range sensor during the mapping stage due to the simplicity of translating the (x, y) coordinates of each image pixel provided by the camera into map coordinates. The perceived environment is represented in a map containing the probability in each cell of the presence of an element/obstacle.

An occupancy grid model is used to represent the environment. As the size of the environment is unknown a priori, it is not possible to create a fixed-size occupancy grid. This is why, the environment is represented as a collection of modular occu-

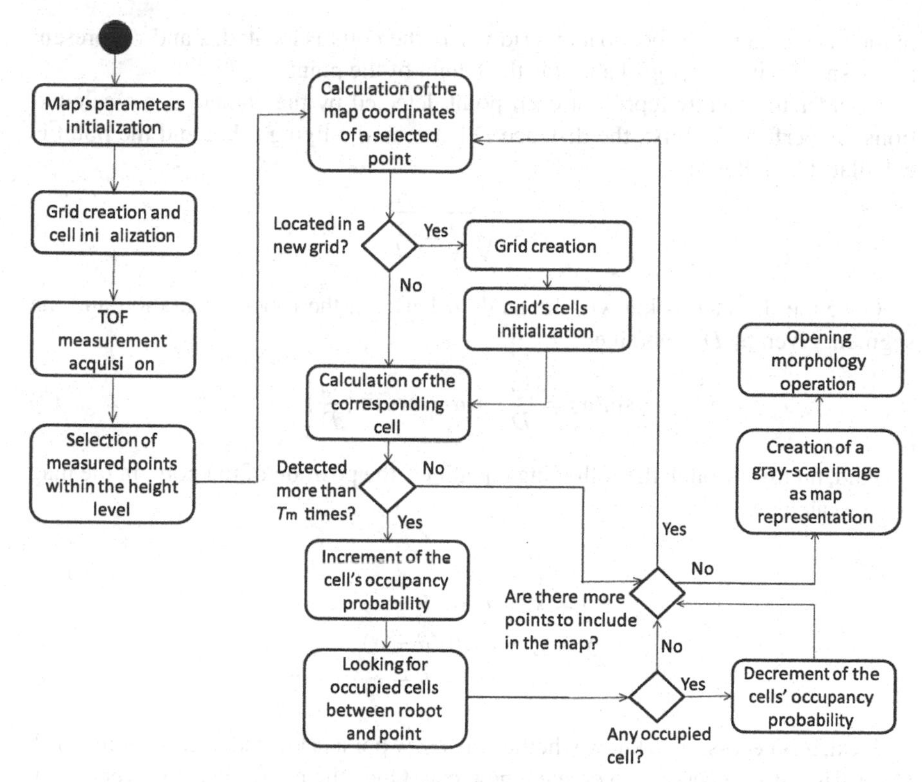

Fig. 1 Flow diagram of the mapping algorithm

pancy grids which are added to the map as far as the robot finds objects outside the existing grids. Therefore, when the robot starts the exploration there exists just one grid, and all necessary grids are added according to the size of the explored environment. Each occupancy grid has the same size and number of cells and is placed in a specific area given by its global coordinates M_x, M_y in the environmental representation. So every point in the environment is located in one and only one grid. For each grid, every cell contains an occupancy value where a value of 0.00 indicates the certainty that the cell is free and a value of 1.00 the certainty that the cell is occupied.

Figure 1 shows a flow diagram of the mapping algorithm. The depth camera provides as output a matrix where each element represents the coordinates (x, y, z) (in meters) of a system. Here the camera is the origin of coordinates, x varies along the horizontal axis, y varies along the vertical axis and z is the distance from the plane defined by the x and y axes. Let us assume that the flying robot position (m_r, x_r, y_r, θ) is known, where m_r is the occupancy grid where the robot is located, x_r and y_r represent the position within that grid, and θ is the robot orientation with respect to the y axis in the grid. On the other hand, the depth camera provides the coordinates of point (x_d, y_d, z_d) as its position relative to the robot's localization. From these two groups of coordinates, it is possible to calculate the position of a point $p(m, x, y, h)$

in the map, being m the occupancy grid where the point is located, x and y represent the position within that grid and h is the height of the point.

In order to visually represent each point detected by the camera some calculations are performed. First, the distance D between the flying robot and the point is calculated as follows:

$$D = \sqrt{x_d^2 + z_d^2} \tag{1}$$

Once the distance is known, the angle α between the robot orientation and the segment given by D is obtained:

$$\sin(\alpha) = \frac{x_d}{D} \rightarrow \alpha = \arcsin(\frac{x_d}{D}) \tag{2}$$

And, finally, through the following equations, the position of the point in the map is also calculated:

$$\begin{aligned} x &= x_r + D\sin(\theta + \alpha) \\ y &= y_r + D\cos(\theta + \alpha) \\ h &= y_d \end{aligned} \tag{3}$$

Next, it is necessary to know whether each new point is located on an existing grid. Otherwise, it is necessary to create a new one. Once the coordinates of every point have been calculated, the new observations must be represented in the grid, updating the occupancy probability of the affected cells. As each cell covers several squared centimeters of the map, it is usual that more than one detected point is located into the same cell for the same observation. Consequently, the occupancy probability of each cell is increased as many times as points are detected to belong to it.

3 Localization from Flying Robot

Only the information provided by the depth camera is used to estimate the flying robot position in the environment. The depth information provided is considered as gray-level values of a traditional image. From these image pixel values, a series of characteristic points (or corners) are extracted to create spatial relations that are placed in the map. When new spatial relations are created from detected corners, they are compared to the previously created ones and identified to calculate their location in the map. Finally the robot position is estimated starting from that information. A flow diagram of the localization algorithm is presented in Fig. 2.

Creating an image representation of the distance measurements enables the use of corner detection algorithm. Concretely, the Harris algorithm has been used for this

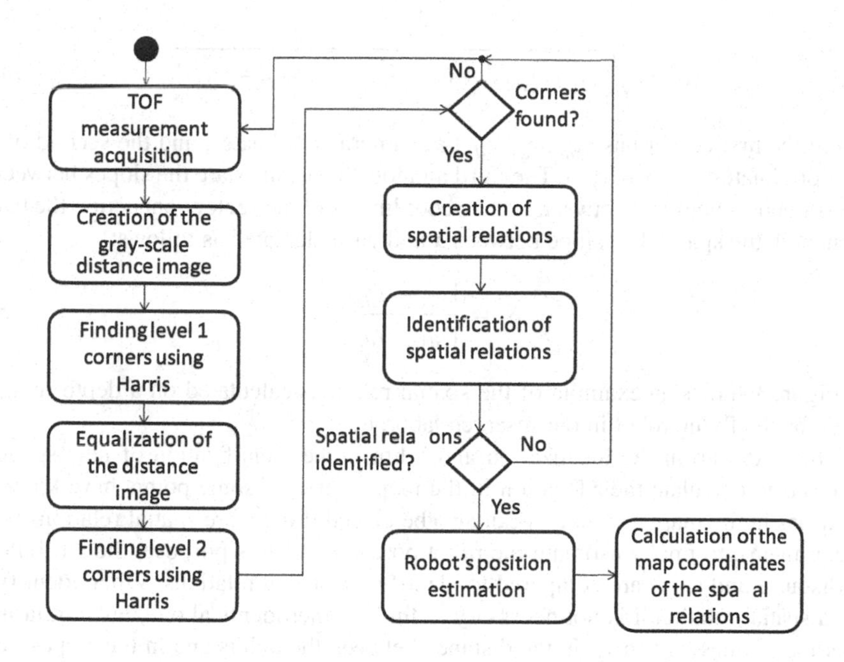

Fig. 2 Flow diagram of the localization algorithm

purpose, mainly because its computational cost is lower than other approaches such as SIFT. Applying Harris algorithm to the distance image results in a list of corners. In our case a hierarchy of two levels of corners is implemented. The first level of corners is composed by those found in the initial distance image, I_d. Usually not many corners are found due to the low contrast of these images, although the gotten corners are quite resistant to noise. Some filters are performed on image I_d to get the second level of corners. A first filter equalizes the image histogram to enhance the contrast. After that, as noise is also enhanced with the equalization, a smoothing Gaussian is applied to the equalized result, obtaining I_f.

Using Harris algorithm on I_f returns more corners than on I_d, but they will be less resistant to noise. Hence, spatial relations based on corners belonging to level 1 are more trusted. So, they have higher priority than those based on corners belonging to level 2. But the last ones are mandatory for a correct localization as level 1 rarely contains enough corners to achieve a good localization.

Once the corners are extracted, spatial relations connecting every pair of points from the same level are established. Some information is associated to these spatial relations in order to define and to identify them. As the map coordinates of the corners are unknown at the beginning, the information associated to spatial relations is independent of the location in the map. The spatial relations information contains the following attributes: priority, according to the level of the corners belonging to the spatial relation, and distance in meters between the two points, N, calculated as follows:

$$N = \sqrt{(x_{d1} - x_{d2})^2 + (y_{d1} - y_{d2})^2 + (z_{d1} - z_{d2})^2} \tag{4}$$

where the first corner has (x_{d1}, y_{d1}, z_{d1}) as camera coordinates, and the second one has coordinates (x_{d2}, y_{d2}, z_{d2}). The third and fourth attributes are the slopes between x and y coordinates and between x and z coordinates of the vector connecting the two corners in the space. The slope between x and y is calculated as follows:

$$S_y = \frac{|x_{d1} - x_{d2}|}{|y_{d1} - y_{d2}|} \tag{5}$$

Figure 3 shows an example of the spatial relations calculated on a depth image taken by the flying robot in our research laboratory.

The next step in the localization algorithm is the identification of new spatial relations to calculate their location in the map (that is, if some points have known positions in the map, new points can also be placed if there are spatial relations between them), to finally estimate the robot position. For this purpose, the attributes of distance and slope are compared to identify the spatial relations. When identifying a spatial relation it is not necessary to find another identical one, but similarity tolerance values τ_{si} and τ_{sl} in the distance between the points and in the slopes, respectively, have been included.

Lastly, the flying robot position can be calculated. Instead of using the coordinates of one point (or corner) like in the mapping algorithm, the coordinates of the two corners that form the spatial relation are used. The system of equations is as follows:

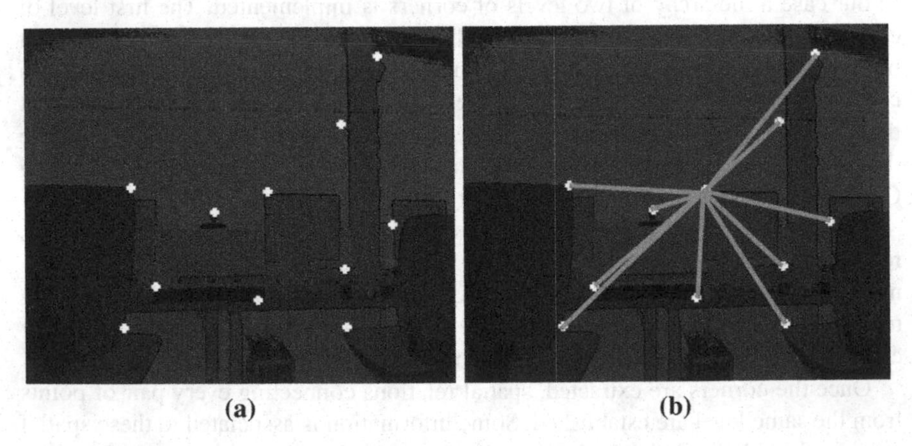

Fig. 3 Example of spatial relations. (a) Depth image with second level corners. (b) Spatial relations corresponding to a random corner

$$
\begin{aligned}
x_r &= x_1 + D_1 \sin(\theta + \pi + \alpha_1) \\
x_r &= x_2 + D_2 \sin(\theta + \pi + \alpha_2) \\
y_r &= y_1 + D_1 \cos(\theta + \pi + \alpha_1) \\
y_r &= y_2 + D_2 \cos(\theta + \pi + \alpha_2)
\end{aligned} \tag{6}
$$

where x_r and y_r are the robot coordinates, θ is the robot orientation, x_1 and y_1 are the coordinates of the first corner, x_2 and y_2 are the coordinates of the second point, D_1 and D_2 are the distances between the robot and each point projected to the floor, and α_1 and α_2 are angles between the flying robot orientation and segments D_1 and D_2, respectively.

From this system of equations, the flying robot orientation is solved as follows:

$$X = x_1 - x_2 \tag{7}$$

$$A = D_1 \cos(\alpha_1) - D_2 \cos(\alpha_2) \tag{8}$$

$$B = D_1 \sin(\alpha_1) - D_2 \sin(\alpha_2) \tag{9}$$

$$\theta = \arccos\left(\frac{-2BX \pm \sqrt{(2BX)^2 - 4(A^2 + B^2)(X^2 - A^2)}}{2(A^2 + B^2)}\right) - \pi \tag{10}$$

After knowing the flying robot orientation, it is possible to calculate the coordinates from the original equation (6).

4 Conclusions

In this paper, the authors have introduced the use of a depth camera for flying robot mapping and localization in home facilities. Firstly, a map building algorithm for flying robots has been introduced. The perceived environment is represented in a map containing in each cell a probability of presence of an object or part of an object. As the size of the environment is unknown a priori, it is not possible to create a fixed-size occupancy grid. The environment is represented as a collection of modular occupancy grids which are added to the map as far as the robot finds objects outside the existing grids.

In our approach the depth camera is exploited as a range sensor for the mapping purpose. Indeed, a depth camera used as a range sensor provides a powerful tool for detecting objects in front of the robot by measuring the distance towards them. Next, for our experiments with localization, the Harris corner detection algorithm is applied. In this case, the depth camera is exploited as a gray-scale camera. The gray-scale image represents distances for the purpose of finding good features to be tracked. These features form the basis of the spatial relations used in the localization algorithm. The approach to the localization problem is based on the computation of the spatial relations existing among the corners detected. The current spatial relations are matched with the relations gotten during previous navigation.

Acknowledgments This work was partially supported by Spanish Ministerio de Economía y Competitividad / FEDER under TIN2013-47074-C2-1-R grant.

References

1. D. Iwakura, K. Nonami, Indoor localization of flying robot by means of infrared sensors. J. Robot. Mechatron. **25**, 201–210 (2013)
2. A. Fernández-Caballero, J.M. Latorre, J.M. Pastor, A. Fernández-Sotos, Improvement of the elderly quality of life and care through smart emotion regulation, *Ambient Assisted Living and Daily Activities*, pp. 348–355, 2014
3. J.C. Castillo, D. Carneiro, J. Serrano-Cuerda, P. Novais, A. Fernández-Caballero, J. Neves, A multi-modal approach for activity classification and fall detection. Int. J. Syst. Sci. **45**, 810–824 (2014)
4. D. Carneiro, J.C. Castillo, P. Novais, A. Fernández-Caballero, J. Neves, Multimodal behavioral analysis for non-invasive stress detection. Expert Syst. Appl. **39**, 13376–13389 (2012)
5. M. Oliver, F. Montero, A. Fernández-Caballero, P. González, J.P. Molina, RGB-D assistive technologies for acquired brain injury: description and assessment of user experience. Expert Syst. (2014). doi:10.1111/exsy.12096
6. A. Briod, P. Kornatowski, J.-C. Zufferey, D. Floreano, A collision-resilient flying robot. J. Field Robot. **31**, 496–509 (2014)
7. K.M. Wurm, C. Stachniss, G. Grisetti, Bridging the gap between feature- and grid-based SLAM. Robot. Auton. Syst. **58**, 140–148 (2010)
8. J. Martínez-Gómez, A. Fernández-Caballero, I. García-Varea, L. Rodríuez, C. Romero-González, A taxonomy of vision systems for ground mobile robots. Int. J. Adv. Robot. Syst. **11**, 111 (2014)
9. T. Collins, Occupancy grid learning using contextual forward modelling. J. Intell. Robot. Syst. **64**, 505–542 (2011)
10. S. Almansa-Valverde, J.C. Castillo, A. Fernández-Caballero, Mobile robot map building from time-of-flight camera. Expert Syst. Appl. **39**, 8835–8843 (2012)
11. A. Ramisa, A. Goldhoorn, D. Aldavert, R. Toledo, R. Lopez de Mantaras, Combining invariant features and the ALV homing method for autonomous robot navigation based on panoramas. J. Intell. Robot. Syst. **64**, 625–649 (2011)
12. M. Bosse, R. Zlot, Keypoint design and evaluation for place recognition in 2D lidar maps. Robot. Auton. Syst. **57**, 1211–1224 (2009)
13. M. Cummins, P. Newman, FAB-MAP: probabilistic localization and mapping in the space of appearance. Int. J. Robot. Res. **27**, 647–665 (2008)
14. D.G. Lowe, Distinctive image features from scale-invariant keypoints. Int. J. Comput. Vis. **60**, 91–110 (2004)
15. G. Arbeiter, J. Fischer, A. Verl, 3D perception and modeling for manipulation on Care-O-Bot 3, in *2010 IEEE International Conference on Robotics and Automation*, p. 5, 2010
16. H. Bay, A. Ess, T. Tuytelaars, L. Van Gool, SURF: speeded up robust features. Comput. Vis. Image Underst. **110**, 346–359 (2008)
17. J. Li, N.M. Allinson, A comprehensive review of current local features for computer vision. Neurocomputing **71**, 1771–1787 (2008)
18. S. Smith, J. Brady, Susan: a new approach to low-level image-processing. Int. J. Comput. Vis. **23**, 45–78 (1997)

19. C. Harris, M. Stephens, A combined corner and edge detector, in *The Fourth Alvey Vision Conference*, pp. 147–151, 1988
20. J. Lee, H. Ko, Gradient-based local affine invariant feature extraction for mobile robot localization in indoor environments. Pattern Recogn. Lett. **29**, 1934–1940 (2008)
21. A. Fernández-Caballero, M.T. López, S. Saiz-Valverde, Dynamic stereoscopic selective visual attention (DSSVA): integrating motion and shape with depth in video segmentation. Expert Syst. Appl. **34**, 1394–1402 (2008)
22. G. Bennett, Probability inequalities for the sum of independent random variables. J. Am. Stat. Assoc. **57**, 33–45 (1962)

Emotion Detection in Ageing Adults from Physiological Sensors

Arturo Martínez-Rodrigo, Roberto Zangróniz, José Manuel Pastor, José Miguel Latorre and Antonio Fernández-Caballero

Abstract The increasing life expectancy is causing a fast ageing population around the globe, which is raising the demand on assistive systems based on ambient intelligence. While numerous papers have focused on the physical aspects in elderly, only a few works have attempted to regulate their emotional state. In this work, a new approach for monitoring and detecting the emotional state in elderly is presented. First, different physiological signals are acquired by means of wearable sensors, and data are transmitted to the embedded system. Next, noise and artifacts are removed by applying different signal processing techniques, depending on the signal behavior. Finally, several temporal and statistical markers are extracted and used to feed the classification model. In this very first version, a logistic regression model is used to detect two possible emotional states. In order to calibrate the model and adjust the boundary decision, twenty volunteers have agreed to be monitored and recorded to train the model. Finally, a decision maker regulates the environment, acting directly upon the elderly's emotional state.

Keywords Emotion detection · Ageing adults · Physiological sensors

A. Martínez-Rodrigo (✉) · R. Zangróniz · J.M. Pastor
Universidad de Castilla-La Mancha, Instituto de Tecnologías Audiovisuales, 16071 Cuenca, Spain

J.M. Latorre
Universidad de Castilla-La Mancha, Instituto de Investigacin En Discapacidades Neurolgicas, 02071 Albacete, Spain

A. Fernández-Caballero
Universidad de Castilla-La Mancha, Instituto de Investigacin En Informitica de Albacete, 02071 Albacete, Spain

A. Mohamed et al. (eds.), *Ambient Intelligence - Software and Applications*, Advances in Intelligent Systems and Computing 376,
DOI 10.1007/978-3-319-19695-4_26

1 Introduction

In the last years, there has been an increase in life expectancy, mostly due to improvements in healthcare systems, progress in medical treatment and growth in long-term care assistance. This has resulted in an increasing ageing population [1]. According to the experts, this trend has a deep impact on economic and social aspects affecting the future quality of life expectation of a large number of people [2]. Additionally, the preference of ageing adults of staying at home is well-known, either due to affective reasons or because the persons suffer from chronic conditions or any kind of disability. However, due to the high cost of institutional living, it is urgent to face this challenge by taking advantage of current assistive technologies. In this regard, a numerous amount of works are making great efforts in developing interactive systems in different areas like healthcare, fitness or entertainment [3]. But less efforts have been made on monitoring and regulating the elderly's emotional state, which is one of the fundamental aspects in self perception of well-being [4]. The main reason lies in the fact that most contemporary human-computer interaction systems are deficient in interpreting emotional information and suffer from the lack of emotional intelligence, i.e. they are unable to identify human emotions to take decisions [5].

A fundamental aspect to build up an emotional assistive system lies in the continuous monitoring of physiological variables during daily activity. Nowadays, recent advances in wireless networks, nanotechnology and integration of micro-controllers in ultra-low power single circuits permit a constant vigilance in a minimally invasive environment for the elderly living conditions [6]. Moreover, wearable sensors are the most appropriate to provide the best detailed user-specific information, since these sensors are placed strategically on the human body or hidden in users' clothes, and are capable of transmitting one or more human physiological signals simultaneously [3]. Furthermore, the capability of reacting upon the ongoing elderly activity is imperative to assemble an emotional system. Thus, such a system should be able of recognising different emotional contexts by using the continuous flow of information provided by the sensors, and, simultaneously, provide personalised services by adapting the reaction to the elderly needs. In this regard, ambient intelligence (AmI) has emerged as a promising approach to face this challenge. Indeed, it takes advantage of pervasive computing, sensor networks and artificial intelligence (AI) [7].

In this work, a new design for emotional state recognition in the elderly is presented. Although an important variety of emotions have been detected by using different frameworks, in this preliminary study only the arousal axis within the arousal-valence space defined by Russell et al. was used [8]. Thus, two different and opposite emotions were quantified, which range from calm (sleepiness) to excited (arousal). The system consists of a physiological signal acquisition module by means of wearables, which communicate with an embedded system capable of detecting these two states in the elderly by using an AmI system, as shown in Fig. 1.

2 Hardware Acquisition

The hardware acquisition layer is a physical layer that includes the wearable systems composed of acquisition circuits and bio-sensors. In this work, the physiological signals are chosen from their capability of reporting the stress degree of a subject, which is indicative of the mental or emotional state [12]. The proposed wearable acquisition device allows long-term and continuous recording of electro-dermal activity (EDA), heart rate (HR), superficial electromyogram (EMG) and skin temperature (SKT). Numerous works have reported their suitability as general indicators of stress [9–11].

The overall system is shown in Fig. 2, where an ultra-low-power, 32-bit ARM Cortex-M3 microcontroller acts as a system control unit. This micro-controller family has been chosen after considering not only the low power consumption but also the scalability into another more powerful Cortex-M microcontroller, capable of handling floating-point operations and signal processing calculations.

Moreover, different acquisition circuits are used to capture physiological variables. Thus, in this scheme, the EDA front-end measures direct current (DC) exosomatic skin conductance through a couple of Ag/AgCl disc electrodes with contact area of $10\,mm^2$.

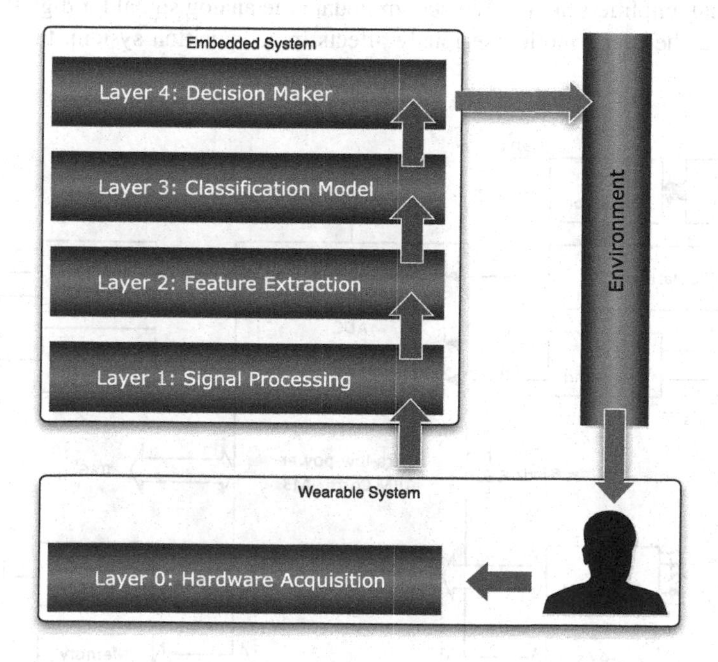

Fig. 1 Levels of abstraction

The electrodes are attached to medial phalanges in palmar sides of left index and middle fingers, since the density of the sweat areas are higher in these locations. A constant-voltage skin conductance circuit is implemented by means of low-noise low-power operational amplifiers. Finally, a 1.5 Hz low-pass filter and 0.05 Hz high-pass filter are responsible for the acquisition of the EDA tonic and phasic components, respectively. It is noteworthy that these analog signals are sampled at 20 Hz throughout 12-bit microcontroller analog-digital converter (ADC). Regarding EMG acquisition, a couple of Ag/AgCl disc electrodes are placed at left forearm extensor digitorum. Next, the differential signal is amplified by means of a low-power high-common-mode rejection instrumental amplifier. In order to increase the signal-to-noise ratio and to reduce the common mode voltage, a third disc is placed at the lateral epicondyle elbow and connected to a driven-right-leg circuit. Finally, the signal is filtered with a low-pass and high-pass filtering of 10 and 400 Hz cut-off frequency, respectively, and subsequently sampled at 1 KHz.

In addition, heart activity is measured by using a commercial photometric front-end, which detects pulse from veins and capillaries in the wrist. This circuit is able to control LED and photo-diode intensity, reject interferences from ambient light, and adapt the signal to be digitalized by the ADC and transmitted to the microcontroller through a serial peripheral interface. Finally, SKT is measured by means of a resistance temperature detector placed on the forearm. A couple of low-noise low-power operational amplifiers are used to accommodate the analog signal for digitalization. Given that the user's motion seriously affects the acquisition system, this way in-

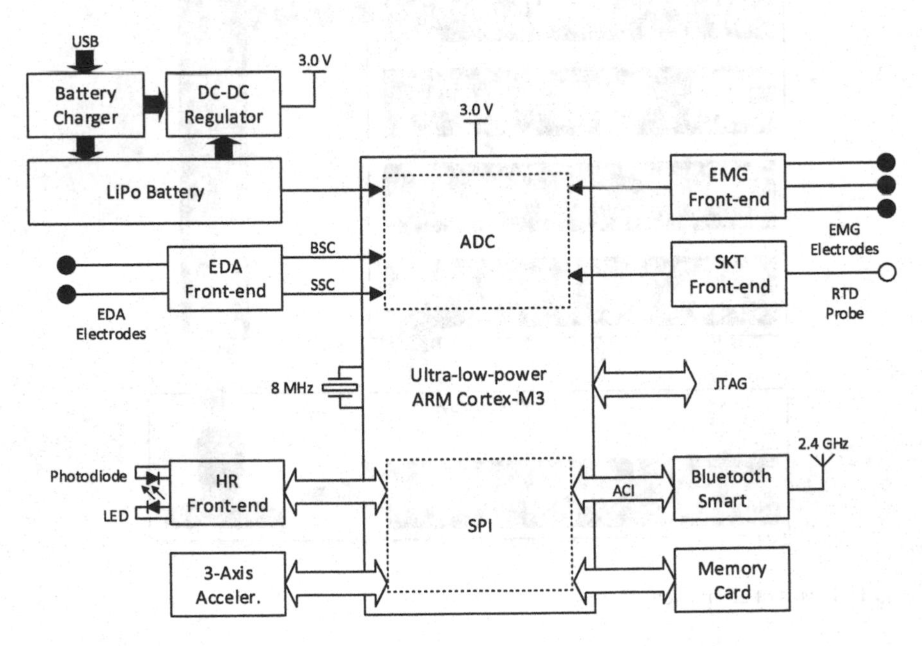

Fig. 2 Hardware scheme

creasing the level of noise and the occurrence of artifacts in the data, a low-power 3-axis accelerometer is embedded in the system to compensate the measurements. The purpose of this sensor is not only to quantify the movement produced in the wearable, but also to register additional features like activity tracker, fall detection, or sleep monitoring.

3 Signal Processing and Feature Extraction

The signal processing layer is responsible for separating useful information to detect physiological events. This layer uses the raw data from the hardware acquisition level, processes the information and bridges to the feature extraction level. In this regard, if processed signals were used directly as input in the classification layer, the efficiency of the model would experience a decreasing. Therefore, the purpose of feature selection is to find out the most representative characteristics of the original data, this way decreasing the quantity of information to be handled by the decision model.

3.1 Electro-Dermal Activity

The EDA signal morphology corresponds to the superposition of two dominant components. On the one hand, the spontaneous skin conductance (SSC) is the result of an increasing activity in the sympathetic nervous system. It is reflected in the signal as a wave with a variable level of amplitude depending of the intensity and duration of the stimulus [13]. On the other hand, the basal skin conductance (BSC) is related both with the sympathetic nervous system and the dermal characteristics of skin [14]. So, it is usually different for each individual.

Considering the different behavior of these two components, BSC is quantified by calculating the mean (BSC_{mean}) over the windowed temporal signal, given that, usually, a slow response can be appreciated in this component. Unlike the basal response, an impulsive signal is shown when a SSC appears, where the amplitude and duration provide valuable information. Therefore, the onset (SSC_{on}) and peak (SSC_{peak}) boundaries are first detected for each event by using an algorithm able to detect sudden slope changes [9]. Then, different markers which evaluate the intensity and duration of the events were computed. These markers have been previously assessed and discriminatory power was reported.(cite DEAP) Thus, SSC_{dur} and SSC_{mag} are computed by evaluating the temporal distance and magnitude differentiation between SSC_{on} and SSC_{peak}, respectively.

3.2 Electromyogram

EMG is the measure of the electrical activation of the muscles, which is controlled by the nervous system and produced during muscle contraction. Unfortunately, several kinds of noise signals and artifacts are usually found when recording an EMG signal. This could mask the EMG activity or report false muscle activation, such an inherent noise in the electrodes, movement artifacts, electromagnetic noise, and cross talk, among others [15]. Consequently, the analysis of EMG with signal processing is imperative to detect muscle activation events and to reduce or eliminate undesired effects provoked by the noise. To this respect, several techniques have been used so far to denoise the EMG signals and amplify the muscle activity, but the discrete wavelet transform (DWT) is still the most popular [15]. Thus, in this work, a DWT is used, applying a Daubechies's function at decomposition level 4, as it had been reported to perform successfully in superficial EMG [16]. Next, the muscle activation event is carried out by applying a simple threshold which determines the existence or absence of EMG activity. Finally, some features involving the muscle activation are calculated. In this regard, the EMG has been previously studied through time, frequency and time-frequency domain. In this work, four temporal markers which involve the EMG integral (EMG_{int}), the mean absolute value (EMG_{mav}), the root mean square (EMG_{rms}) and the variance (EMG_{var}) of the detected EMG events are used. These markers have been previously used and discriminatory power was reported [17].

3.3 Heart Activity

Electrical activation of the heart takes place by means of the activation of the sinoatrial node which spreads out an electrical impulse from the atrium to ventricles. Heart activity is reflected in the electrocardiogram (ECG) through the P-wave, QRS complex and T-wave, which represents the atrial depolarisation, the ventricular polarization and ventricular depolarisation, respectively. In this regard, while P-wave usually appears in the ECG as a noisy and low amplitude wave, the QRS complex presents the highest amplitude within the ECG, and more specifically the R peak. For this reason, R peak is usually designed to locate the heart cycle within continuous heart activity. In order to track this activity, the R peak is located at each heart cycle and, then, the distance between consecutive R peaks are measured (RR series). This way, the heart rhythm or heart rate is obtained. Nevertheless, given that heart activity is caused by the autonomous nervous system, heart rate variability (HRV), showing the alterations of heart rhythm, is usually computed to evaluate the arousal level of an individual. Although HRV has been analysed from different points of view, in this preliminary work only a few temporal markers are calculated. Thus, the R peaks are firstly located within the ECG by using a second-order derivative algorithm. Then, several temporal features are calculated over the RR-intervals including mean (RR_{mean}), standard deviation (RR_{std}) and root mean square (RR_{rms}).

3.4 Temperature

When human body is under stress, muscles are tensioned, contracting the blood vessels and thus provoking a decrease of temperature [19]. Unlike the aforementioned features, SKT changes are relatively low, such that long-term variability monitoring is necessary. In this work, a mean of temperature, considering a five-minutes temporal window, is used to compute SKT_{mean}.

4 Classification Model and Decision Maker

In order to achieve an adequate system response, a massive quantity of information from the sensors needs to be organised and subsequently used to make decisions. Taking decisions implies an information model, which entails hierarchical levels formed by layers which range from data acquisition to decision making, as shown in Fig. 1.

4.1 Learning Algorithm

A preliminary learning process is necessary to compute the decision boundary in order to classify the arousal level in the elderly. In this sense, twenty volunteers were enrolled and agreed to visualise a set of fifty images each one to perform an experiment. The set is composed of two subsets of pictures extracted from the International Affective Picture System [20]. In this regard, 25 out of 50 samples correspond to pictures which report a high arousal level (HAL), and the other 25 out of 50 samples correspond to pictures which report a low arousal level (LAL) [20], thus gathering a total of 1000 samples (n = 1000). Each person is monitored and physiological data for each sensor are recorded. Once the data has been processed, the physiological markers are extracted and used to feed the learning algorithm. In this sense, eleven markers (m = 11) described in Sect. 2 are computed for each person and picture. The feature matrix is calculated as shown in Eq. 1.

$$X_{m,n} = \begin{vmatrix} x_{(1,1)} & x_{(1,2)} & \cdots & x_{(1,n-1)} & x_{(1,n)} \\ x_{(2,1)} & x_{(2,2)} & \cdots & x_{(2,n-1)} & x_{(2,n)} \\ \vdots & \vdots & \ddots & \vdots & \vdots \\ x_{(m,1)} & x_{(m,2)} & \cdots & x_{(m,n-1)} & x_{(m,n)} \end{vmatrix} \quad (1)$$

where $x_{m,n}$ corresponds to the n-th sample for the m-th feature. Also, given the two possible outputs, y_n is mapped such that $y_n = 0$ if the n-th sample is labeled as LAL and $y_n = 1$ if the n-th sample is labeled as HAL. Then, the cost function is computed as:

$$C_\theta = \frac{1}{m} \sum_{i=1}^{m} [-y_i log(h_\theta(x_{(i)})) - (1 - y_{(i)}) log(1 - h_\theta(x_{(i)}))] \quad (2)$$

where h_θ corresponds to the logistic regression hypothesis defined throughout the sigmoid function (s), such that:

$$h_\theta(x) = s(\theta^T x) \quad (3)$$

Finally, the optimal parameters of the logistic regression model (θ) are estimated by minimising the cost function C. Thus, the decision boundary, which forms the decision maker layer, is calculated.

5 Conclusions

In this work, a method for evaluation the emotional state in elderly is presented. First, the signal acquisition was accomplished by designing a new hardware able to recording different physiological signals. Next, signal processing and feature extraction were performed to feed the classification system. A very first version of classification system was presented here, where only two possible outcomes were considered, given the arousal level of the users. Futures lines of researching will focus on increasing the classification system complexity, in order to take into consideration a wider range of emotions.

Acknowledgments This work was partially supported by Spanish Ministerio de Economía y Competitividad / FEDER under TIN2013-47074-C2-1-R grant.

References

1. World Health Organization, in *Ageing and Life Course* (2011)
2. S. Mowafey, S. Gardner, A novel adaptive approach for home care ambient intelligent environments with an emotion-aware system, in *UKACC International Conference on Control* (Cardiff, UK, 2012), pp. 771–777, 3–5 Sept 2012
3. M.A. Hanson, H.C. Powell Jr., A.T. Barth, K. Ringgenberg, B.H. Calhoun, J.H. Aylor, J. Lach, Body area sensors networks: challenges and opportunities, in *IEEE Computer Society*, pp. 58–65, 2009
4. A. Fernández-Caballero, J.M. Latorre, J.M. Pastor, A. Fernández-Sotos, Improvement of the elderly quality of life and care through smart emotion regulation, in *Ambient Assisted Living and Daily Activities*, pp. 348–355, 2014

5. S. Koelstra, C. Muhl, M. Soleymani, J.-S. Lee, A. Yazdani, T. Ebrahimi, T. Pun, A. Nijholt, I. Patras, DEAP: a database for emotion analysis using physiological signals. IEEE Trans. Affect. Comput. **3**(1), 18–31 (2012)
6. M. Chen, S. Gonzalez, A. Vasilakos, H. Cao, V.C.M. Leung, Body area networks: a survey. Mobile Netw. Appl. **16**, 171–193 (2011)
7. P. Remagnino, G.L. Foresti, Ambient intelligence: a new multidisciplinary paradigm. IEEE Trans. Syst. Man Cybern. Part A **35**(1), 1–6 (2005)
8. J.A. Russell, A circumplex model of affect. J. Pers. Soc. Psychol. **39**(6), 1161–1178 (1980)
9. J.A. Healey, R.W. Picard, Detecting stress during real-world driving tasks using physiological sensors. IEEE Trans. Intell. Trans. Syst. **6**(2), 156–166 (2005)
10. J.A. Veltman, A.W.K. Gaillard, Physiological indicies of workload in a simulated flight task. Biol. Psychol. **42**, 323–342 (1996)
11. K. Nagamine, A. Nozawa, H. Ide, Evaluation of emotions by Nasal Skin temperature on auditory stimulus and olfactory stimulus. IEEJ Trans. EIS **124**(9), 1914–1915 (2004)
12. J. Herbert, Fortnightly review: stress, the brain, and mental illness. British Med. J. **315**, 530–535 (1997)
13. L. Lidberg, G. Wallin, Sympathhetic skin nerve discharges in relation to amplitude of skin resistance responses. Psychopysiology **18**(3), 268–270 (1981)
14. P.H. Venables, M.J. Christie, Electrodermal activity, *Techniques in, Psychophysiology* (2012)
15. R. Chowdhury, M. Reaz, A.M. Mohd, A. Bakar, K. Chellappan, T. Chang, Surface electromyography signal processing and classification techniques. Sensors **13**, 12431–12466 (2013)
16. G. Wei, F. Tian, G. Tang, C. Wang, A wavelet-based method to predict muscle forces from surface electromyography signals in weightlifting. J. Bionic Eng. **9**, 48–58 (2012)
17. B. Hudgins, P. Parker, R.N. Scott, A new strategy for multifunction myoelectric control. IEEE Trans. Biomed. Eng. **40**, 8294 (1993)
18. M. Malik, J.T. Bigger, A.J. Camm, R.E. Kleiger, A. Malliani, A.J. Moss, P.J. Schwartz, Heart rate variability: standards of measurement, physiological interpretation, and clinical use. Eur. Heart J. **17**(2), 1043–1065 (1996)
19. P. Leijdekkers, V. Gay, W. Frederick, CaptureMyEmotion: a mobile app to improve emotion learning for autistic children using sensors, in *26th IEEE International Symposium on Computer-Based Medical Systems*, pp. 381–384, 2013
20. P.J. Lang, M.M. Bradley, B.N. Cuthbert, *International Affective Picture System (IAPS): Affective Ratings of Pictures and Instruction Manual* (Technical Report A-8. University of Florida, Gainesville, 2008)

Augmented Tangible Surfaces to Support Cognitive Games for Ageing People

Fernando Garcia-Sanjuan, Javier Jaen and Alejandro Catala

Abstract The continuous and rapidly increasing elderly population requires a revision of technology design in order to devise systems usable and meaningful for this social group. Most applications for ageing people are built to provide supporting services, taking into account the physical and cognitive abilities that decrease over time. However, this paper focuses on building technology to improve such capacities, or at least slow down their decline, through cognitive games. This is achieved by means of a digitally-augmented table-like surface that combines touch with tangible input for a more natural, intuitive, and appealing means of interaction. Its construction materials make it an affordable device likely to be used in retirement homes in the context of therapeutic activities, and its form factor enables a versatile, quick, and scalable configuration, as well as a socializing experience.

Keywords Gerontechnology · Cognitive games · Socialization · Collaboration · Tangible user interfaces (TUI) · Surfaces

1 Introduction

The number of ageing people in the European Union is fiercely increasing. According to Eurostat's statistics,[1] EU's elderly population is expected to rise from 17.9 % in 2012 to 28.1 % by the year 2050 due to the average increase of life

[1]http://ec.europa.eu/eurostat/statistics-explained/index.php/Population_structure_and_ageing.

F. Garcia-Sanjuan (✉) · J. Jaen · A. Catala
Grupo ISSI, Departamento de Sistemas Informáticos Y Computación,
Universitat Politècnica de València, Camí de Vera S/N, 46022 Valencia, Spain
e-mail: fegarcia@dsic.upv.es

J. Jaen
e-mail: fjaen@dsic.upv.es

A. Catala
e-mail: acatala@dsic.upv.es

A. Mohamed et al. (eds.), *Ambient Intelligence - Software and Applications*,
Advances in Intelligent Systems and Computing 376,
DOI 10.1007/978-3-319-19695-4_27

expectancy and the continuous decline in birth rates. This growth will require adapting existing technological services and creating new ones for this group of people [1].

The idea of ageing people and technology being incompatible is a cliché, as it has been already proven in the literature. It is not true that the elderly have not the capacity or the will to learn and use new technologies. They do have the ability, although not necessarily the necessity [2]. It would appear that, traditionally, technological devices have been designed for youngsters, and neither their purpose nor interfaces are appealing to ageing people. In fact, a study conducted by Fisk et al. [3] concluded that more than half of the problems that elders experience with technology were associated with usability issues. In particular, the design of input/output devices and user interfaces is critical because they interact with the user's perceptual and sensorial system, which, at certain age, experience some changes that may have a negative impact on usability [3, 4]. Examples of these changes are decrease of visual and acoustic capacities, touch- and movement-related issues (such as arthritis, tremors, walking problems, etc.), and a reduction of some cognitive capacities [5].

Traditionally, the most common ways of interacting with computers were using mouse and keyboard, but these present severe usability issues that can cause the elderly to be reluctant to engage with technology [6]. Direct contact via touch interfaces, instead, has shown to be more adequate to ageing users since these interfaces present less cognitive load and less spatial demand, and many efforts are being made as of late in order to create more intuitive user experiences using this kind of input devices [7]. Furthermore, Torres [6] proposes to devise alternative ways of performing input, for example, via tangible interfaces, which are typically referred as Tangible User Interfaces (TUIs) [8]. These offer spatial mapping, input/output unification, and the support of trial-and-error actions that can exploit innate spatial and tactile abilities; and have already been used successfully in cognitive training activities [9].

The present work contributes to the field with a TUI prototype in the form of a table-like surface aiming at building games for the elderly to train their cognitive abilities (see Fig. 1). It intends to be usable by providing a scalable and versatile means of configuration for both ageing people and the therapists who design the games, and by enabling a more natural interaction through tangible manipulations along with fully supporting touch interactions. Another important purpose of our proposed infrastructure is to foster socialization and the training of cognitive abilities that can improve elders' quality of life. The rest of the document is structured as follows: First, related work on technology for the elderly is described; then, our augmented tangible surface is presented; Sect. 4 explains how our prototype could be used to build games for elders' cognitive training; and, finally, future work and conclusions are drawn.

Fig. 1 Example of a game running on the surface

2 Related Work

Many research works have proposed technology to help ageing people deal with age-related problems in respect of physical and cognitive capacities. Some of them, in the form of assistive robots or mobile applications (e.g., [10–12]), offer services that improve the quality of life of the elderly and enhance their independence. However, they are often devised as aiding tools and not as therapeutic mechanisms to reduce the negative impact of their decreasing capacities. Besides, they are usually designed as private devices, omitting socialization, despite ageing people seem to assign a high value to socialization and they even report being against technology when it replaces face-to-face interactions [13]. In terms of socialization, there have been efforts using robots, called assistive social robots [14], focusing on improving socialization between themselves and ageing users. However, they do not intend to foster human-to-human socialization. In fact, some authors have expressed their concerns about these technologies incrementing social isolation [15].

In addition to robots, other works propose the use of digital games (a.k.a. cognitive games) that stimulate the previously mentioned decreasing cognitive abilities, and also foster socialization (e.g., [4, 16, 17]). In this sense, playing represents an advantageous way to engage elder users both cognitively and socially [5]. There are many references in the literature stressing the benefits of playing videogames for the elderly. They have been proved to decrease reaction times [18, 19], and improve quality of life, self-confidence, and cognitive skills (these two showing a positive correlation) [6]. Moreover, Whitcomb [16] analyzed how ageing people played a series of videogames, and observed that they increased social interaction and perceptual-motor capacities (eye-hand coordination, dexterity, fine motor ability, and a reduction of the reaction time). Also, although the author did not explicitly study how videogames affected cognitive capacities, the study detected a positive effect of videogames on information processing, reading, comprehension, and memory.

Interaction design for videogames targeting elder people is a critical dimension to be considered. Whitcomb [16] also enumerates several characteristics that make a videogame unsuitable for them, such as small-sized objects, rapid movements or reactions required, or the sound being inappropriate. In terms of interaction mechanisms, this study focused on computer games, which are mainly interacted through mouse and keyboard. However, other interaction mechanisms, as we discussed in the previous section, may be advantageous when considering ageing people. In this respect, authors such as Jung et al. [20] have explored other input/output devices, e.g. a Wii stick in a game to enhance general wellbeing (physical activity, self-esteem, affect, and level of solitude). However, in our opinion, this type of interaction should be considered with caution when the elderly are involved because it has been reported to produce physical lesions such as tendinitis (or Wiiitis as it has been called) [21]. Alternatively, Chiang et al. [22], through Kinect games, report elder users improving significantly their visual performance skills. Others, however, have taken advantage of the increasing popularization of handheld devices. MemoryLane [4], although not exactly a game, fosters reminiscence through a PDA application to create "memory stories" with pictures. Vasconcelos et al. present CogniPlay [17], a gaming platform running on tablets which includes several games to stimulate cognitive abilities, such as matching pairs to enhance short-term memory, and social interaction through competition. However, the consideration of small devices or elements that are designed to be used by a single user is clearly a step in the wrong direction when collaboration needs to be fostered.

Our approach intends to merge touch interaction capabilities provided by handheld devices such as tablets but at the same time taking advantage of the natural and intuitive manipulations that physical (tangible) elements can bring. Moreover, by proposing a surface-like configuration with such affordable materials and devices, a cost-effective public space can be built where elder players can all have a simultaneous and equal access to the game space which fosters collaboration.

3 Designing the Prototype

The prototype presented in this work aims at supporting collaborative therapeutic games for the elderly around physical tables. Current digital tabletop technology would indeed allow us to deliver fine-grained touch interactions and high-end visual representations, but fully interactive tabletops are still expensive, and their form factor complicates their mobility and scalability. Instead, we propose a cost-effective way of creating a surface by arranging physical tiles, which can form a table-like interactive 2D environment of any arbitrary topology anywhere on a flat terrain. The resulting surface becomes a public space where all users can collaborate in problem solving tasks, and therapists can design cognitive games, such as matching pairs to train short-term memory, simply by handling the physical tiles, without any technological knowledge required. This type of surface is digitally augmented in order to provide richer features to the games. However, to decrease the decoupling between the physical and the digital space that would take place if the digital information was shown in a separate display held by the players, a tablet is instead mounted on a small mobile robot that moves through the physical surface displaying digital contents within the context of the physical space (see Fig. 1). With respect to the input mechanisms supported for the elder participants, they can both use touch contacts and gestures on the tablet, and interact directly with the physical surface by adding and removing tiles that have a specific digital behavior associated or by giving commands to the robot using special physical tags, hence providing a more natural and intuitive interaction.

Our augmented tangible surface consists of two major components, as can be seen in Fig. 2. The 2D surface can be constructed by arranging several 20 cm × 20 cm tiles following any desired flat configuration. Each tile has a number of black lines which allow the robot to move in the physical space in different directions by following them. The lines may represent a crossroads for the robot to choose which direction to take, or a specific path such as a curve. Depending on the game, each tile may also contain some drawings that make sense to the users in the context of the activity being developed. As Fig. 3 (left) depicts, each tile consists of a squared piece of paper with the path black lines and possibly the drawings, an RFID tag to be read by the robot when passes over it and which provides the tablet with digital information, another piece of paper with only the path printed (representing the back of the tile), and two pieces of plastic to protect it all.

The robot has been constructed using Lego™ Mindstorms® Ev3 and it has an Android tablet on it that serves as a rich colorful digital input/output device. Figure 3 (right) shows the different components this robot is composed of (aside from the tablet). It has a color sensor that differentiates between black and white so it can follow the surface's black paths. Every time it reaches the center of a tile, its RFID reader situated on the bottom reads the tag embedded in the tile and sends its code to the tablet. This one contains the game logic, handles touch interaction, and sends the proper control commands to the robot via Bluetooth.

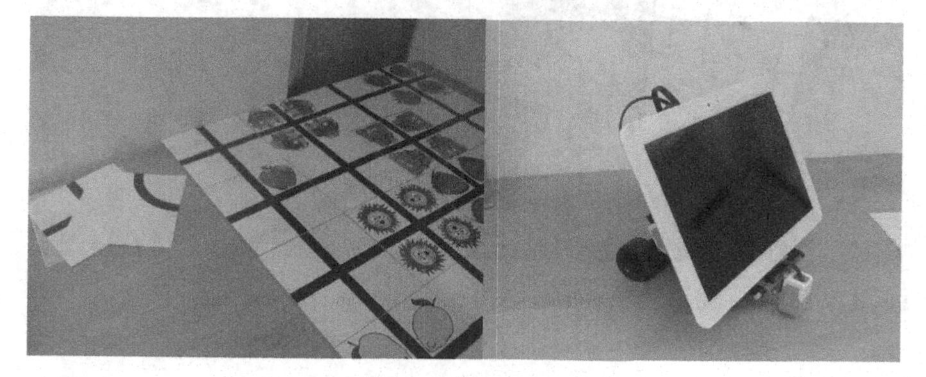

Fig. 2 The tangible surface's parts. On the left, the different tiles that compose the physical surface. On the right, the mobile robot that displays the digital content

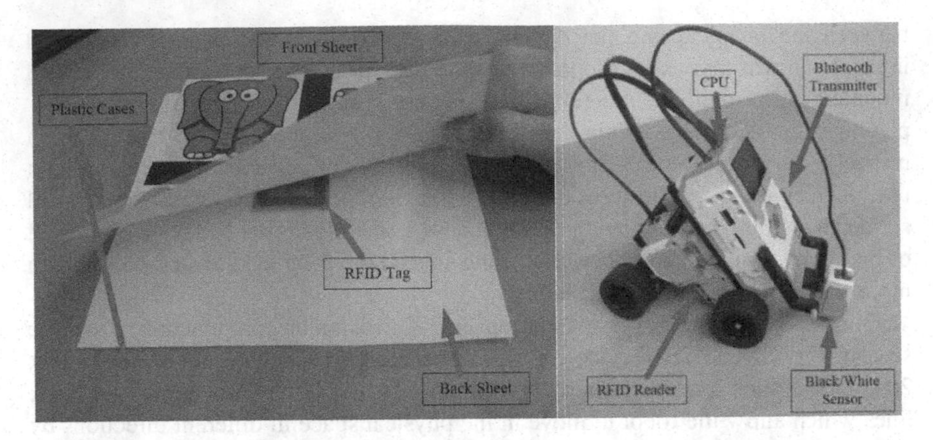

Fig. 3 Details of the tile (left) and the mobile robot (right)

The system allows several interaction modalities: Users can perform coarse-grained interactions by coupling and decoupling the tiles at will at runtime or by using command cards that are read by the robot's RFID reader. On the other hand, finer-grained interactions can also be achieved via touch contacts on the tablet. Since different tiles have distinct RFID codes, they can provide the game with different information, thereby removing the need of touch input, and leaving the tablet for display purposes only if this would be required. Figure 4 shows an example of a touch interaction (left), where the user touches the tablet designing a path for the robot to follow, and of a tangible interaction (right), where the player physically "draws" the path in the surface by rearranging the tiles.

Fig. 4 Interaction modalities supported. Touch-based (left) and tangible (right)

4 A Game to Stimulate Cognitive Abilities for the Elderly

Ageing entails a diminution in some physical and cognitive capacities. Examples of these reduced capacities are short-term (working) memory, the ability to filter irrelevant information, divided attention, and visual-spatial attention [5]. This section exemplifies how the prototype described in this paper can be used to help training these capacities and fight their decline through developing cognitive games. An illustrative scenario of a game to improve short-term memory and divided attention is detailed next.

A therapist arranges several tiles containing pictures on the surface as depicted in Fig. 4 for the player to memorize. Then, the former turns over the tiles, removing the pictures and leaving only the crossing paths visible. At this moment, the tablet shows a target image and the user must devise a path from the robot position to the location where the displayed figure is. Then, using either touch or tangible interaction, the users draw the trajectory and the robot follows it. The system provides positive or negative feedback depending on the adequacy of the path defined to reach the proposed target. In additional iterations of the game the therapist also includes tiles showing a wrong direction sign to motivate the users to find alternative paths to the target avoiding these tiles.

Following Salthouse and Babcock's suggestions [23], both the speed of the robot and the rate at which the target elements are displayed should be reduced. According to Rogers [24], the reduction of the ability to filter irrelevant information affects the selective attention, which depends on the familiarity of the user with the presented objects. Taking this into account, the images to memorize should be easily recognizable by the participants. This game requires using divided attention because players must focus on remembering the location of the images and drawing paths at the same time. This training by itself, as stated by Rogers [24] enhances divided attention and improves making attention switches. Also, the remembrance of the objects' location and the creation of routes serve as a training of visual-spatial processes.

Another important aim of this platform is avoiding the dangers of social isolation that could provide a similar implementation where each user would hold a tablet. The intrinsic nature of our table-like surface enables collaborative scenarios where several people situate around the physical table and help one another find better pathways and/or simply discuss the game and the situation themselves.

5 Conclusions and Future Work

In this paper we presented a prototype of a digitally-augmented tangible surface aimed at constructing cognitive games for the elderly. Not only the table-like design fosters human-to-human socialization via collaboration but also the touch and tangible capabilities bring more natural and intuitive interactions that can appeal ageing users. Hence, the ultimate purpose of the present work is to design useful and usable technology for this special group.

The platform is built with cost-effective materials, and its design allows for a quick setup and a high versatility and scalability. We exemplified the use of the surface with a cognitive game to improve short-term memory and selective, divided, and visual-spatial attention.

As future work, we intend to perform experiments with real users in order to test whether the tangible interaction offers any advantages with respect to digital (touch) both in configuring the layout of the game (i.e., the arrangement of the tiles) and in the problem solving phase (e.g., drawing a path for the robot to follow or giving it specific instructions at run time). Other future experiments will focus on the actual perceived usefulness and usability of the platform and on whether this system has any positive effect on the already enumerated cognitive capacities meant to be stimulated.

Acknowledgments This work received financial support from Spanish Ministry of Economy and Competitiveness under the National Strategic Program of Research and Project TIN2010-20488, and from Universitat Politècnica de València under Project UPV-FE-2014-24. It is also supported by fellowships APOSTD/2013/013 and ACIF/2014/214 within the VALi+d program from Conselleria d'Educació, Cultura i Esport (GVA).

References

1. F. Nunes, P.A. Silva, F. Abrantes, Human-computer interaction and the older adult: an example using user research and personas, in *Proceedings of PETRA '10* (ACM, New York, 2010), pp. 49:1–49:8
2. J. Durick, T. Robertson, M. Brereton, F. Vetere, B. Nansen, Dispelling ageing myths in technology design, in *Proceedings of OzCHI '13* (ACM, New York, 2013), pp. 467–476
3. A.D. Fisk, W.A. Rogers, N. Charness, S.J. Czaja, J. Sharit, *Designing for Older Adults: Principles and Creative Human Factors Approaches* (CRC Press, 2004)
4. S.M. Carthy, H. Sayers, P.M. Kevitt, M. McTear, MemoryLane: reminiscence for older adults, in *Proceedings of 1st International Workshop on Reminiscence Systems*, pp. 22–27, 2009
5. L. Gamberini, M. Alcaniz, G. Barresi, M. Fabregat, F. Ibanez, L. Prontu, Cognition, technology and games for the elderly: an introduction to ELDERGAMES project. PsychNol. J. **4**, 285–308 (2006)
6. A.C.S. Torres, Cognitive effects of video games on old people. Int. J. Disabil. Hum. Dev. **10**, 55–58 (2011)
7. B. Loureiro, R. Rodrigues, Multi-touch as a natural user interface for elders: a survey, in *6th Iberian Conference on Information Systems and Technologies* (IEEE, 2011), pp. 1–6
8. H. Ishii, B. Ullmer, Tangible bits: towards seamless interfaces between people, bits and atoms, in *Proceedings of CHI '97* (ACM, New York, 1997), pp. 234–241
9. E. Sharlin, Y. Itoh, B. Watson, Y. Kitamura, S. Sutphen, L. Liu, F. Kishino, Spatial tangible user interfaces for cognitive assessment and training, in *Biologically Inspired Approaches to Advanced Information Technology*, ed. by A.J. Ijspeert, M. Murata, N. Wakamiya (Springer, Berlin Heidelberg, 2004), pp. 137–152
10. M. Montemerlo, J. Pineau, N. Roy, S. Thrun, V. Verma, Experiences with a mobile robotic guide for the elderly, in *Proceedings of AAAI '02* (AAAI, Menlo Park, 2002), pp. 587–592

11. B. Otjacques, M. Krier, F. Feltz, D. Ferring, M. Hoffmann, Helping older people to manage their social activities at the retirement home, in *Proceedings of BCS-HCI '09* (BCS, Swinton, 2009), pp. 375–380
12. J. Goodman, S. Brewster, P. Gray, Older people, mobile devices and navigation, in *Proceedings of HCI '04, HCI and the Older Population Workshop*, pp. 13–14, 2004
13. S. Eggermont, H. Vandebosch, S. Steyaert, Towards the desired future of the elderly and ICT: policy recommendations based on a dialogue with senior citizens. Poiesis Prax. **4**, 199–217 (2006)
14. J. Broekens, M. Heerink, H. Rosendal, Assistive social robots in elderly care: a review. Gerontechnology **8**, 94–103 (2009)
15. A. Sharkey, N. Sharkey, Granny and the robots: ethical issues in robot care for the elderly. Ethics Inf. Technol. **14**, 27–40 (2012)
16. G.R. Whitcomb, Computer games for the elderly, in *Proceedings of CQL '90* (ACM, New York, 1990), pp. 112–115
17. A. Vasconcelos, P.A. Silva, J. Caseiro, F. Nunes, L.F. Teixeira, Designing tablet-based games for seniors: the example of cogniplay, a cognitive gaming platform, in *Proceedings of FnG '12* (ACM, New York, 2012), pp. 1–10
18. J.E. Clark, A.K. Lanphear, C.C. Riddick, The effects of videogame playing on the response selection processing of elderly adults. J. Gerontol. **42**, 82–85 (1987)
19. R.E. Dustman, R.Y. Emmerson, L.A. Steinhaus, D.E. Shearer, T.J. Dustman, The effects of videogame playing on neuropsychological performance of elderly individuals. J. Gerontol. **47**, 168–171 (1992)
20. Y. Jung, K.J. Li, N.S. Janissa, W.L.C. Gladys, K.M. Lee, Games for a better life: effects of playing Wii games on the well-being of seniors in a long-term care facility, in *Proceedings of IE '09* (ACM, New York, 2009), pp. 5:1–5:6
21. J. Bonis, Acute Wiiitis. N. Engl. J. Med. **353**, 2431–2432 (2007)
22. I.-T. Chiang, J.-C. Tsai, S.-T. Chen, Using Xbox 360 Kinect games on enhancing visual performance skills on institutionalized older adults with wheelchairs, in *Proceedings of DIGITEL '12* (IEEE, 2012), pp. 263–267
23. T.A. Salthouse, R.L. Babcock, Decomposing adult age differences in working memory. Dev. Psychol. **27**, 763–776 (1991)
24. W.A. Rogers, Attention and aging, in *Cognitive aging: a primer*, ed. by D.C. Park, N. Schwarz (Psychology Press, New York, 2000), pp. 57–73

Author Index

A. Mohamed et al. (eds.), *Ambient Intelligence - Software and Applications*,
Advances in Intelligent Systems and Computing 376,
DOI 10.1007/978-3-319-19695-4

Printed in the United States
By Bookmasters